W9-BMT-747

THE
GARDENER'S
ATLAS

A FIREFLY BOOK

Published in Canada in 1998 by
Firefly Books Ltd.
3680 Victoria Park Avenue
Willowdale, Ontario
M2H 3K1

Published in the United States
in 1998 by
Firefly Books (U.S.) Inc.
P.O. Box 1338, Ellicott Station
Buffalo, New York
14205

Cataloguing in Publication Data
Grimshaw, John Michael 1967-
The gardener's atlas
Includes index.
ISBN 1-55209-226-7
1. Flowers—History. 2. Flower
gardening. I. Title.
SB404.5.G74 1998 635.9
C98-930751-4

Project editor:
Gilly Cameron Cooper

Copy editors:
Gilly Cameron Cooper,
Sally MacEachern,
Maggi McCormick, Steve Parker

Design: Steve McCurdy
Cartographer: Julian Baker
Picture research: Gill Metcalfe
Illustrator: Kevin Maddison
Art editor: Elizabeth Healey

Color separation by Bright Arts
(Singapore) Pte. Ltd.
Printed and bound by Star
Standard Industries (Pte.) Ltd.,
Singapore

THE
GARDENER'S
ATLAS

THE ORIGINS, DISCOVERY, AND CULTIVATION
OF THE WORLD'S MOST POPULAR GARDEN PLANTS

DR. JOHN GRIMSHAW
Consultant
DR. BOBBY WARD

FIREFLY BOOKS

CONTENTS

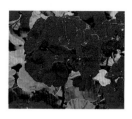

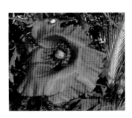

EVERY GARDEN, HOWEVER SMALL,
CONTAINS PLANTS WITH A PAST.
LEARNING THEIR STORIES ADDS RICHNESS TO THE
GARDENING EXPERIENCE.

INTRODUCTION

Over 4,000 years of tradition, science, and art are at the disposal of today's gardeners. The plants in your garden may be native stock, exotics with origins on the other side of the world, or cultivars that are the product of centuries of careful attention by gardeners who have selected better clones from wild populations, or who have laboriously bred them to conform to their ideal of perfection. This book tells the story of these plants and of gardens and gardeners from the earliest records to the present day. The labors of plant hunters and the efforts of plant breeders have made available plants suited to every situation, and even a small plot can grow some plant with a fascinating history to tell.

Although, in most urbanized countries, the size of gardens is shrinking, gardeners can still draw inspiration from the past, adapting traditions to create their own miniature paradise. The great parterres (formal, intricately patterned flowerbeds) of the palace of Versailles may be too large for your plot, and the simple beauty of the gardens of the Alhambra may be out of place in suburbia. But the imaginative gardener can learn from both. Flowers, or even vegetables, can be attractively grown in box-edged, parterre-like beds, while a pool and miniature fountain can create the ambience of a Moorish garden.

PRIVATE PARADISE
A plantsman's garden in
England combines plants
from all over the world
with a strong design,
creating a private paradise
that embodies 4,000 years
of horticultural tradition.

*"*GOD ALMIGHTY FIRST PLANTED A GARDEN. AND
INDEED, IT IS THE PUREST OF HUMAN PLEASURES. IT IS
THE GREATEST REFRESHMENT TO THE SPIRITS OF MAN"—FRANCIS BACON
(1561-1626) WROTE:

PARADISE FOUND

The practice of cultivating plants for their decorative value seems to have first arisen in the "fertile crescent" of Mesopotamia, between the rivers Tigris and Euphrates in modern Iraq. The gardens and orchards of Sumer in the third millennium B.C. were praised in the *Epic of Gilgamesh*; later records show that the Assyrians extended the horticultural traditions of Mesopotamia. They created private gardens, as well as large parks planted with ornamental trees and stocked with wild animals. These were known in Ancient Greek as *paradeisoi*—the original paradise gardens, whose imagery was to inspire the Biblical Garden of Eden, and influence the religious traditions of the world.

Ideal landscapes

The most famous early recreations of an ideal landscape were the Hanging Gardens of Babylon, considered to be among the Seven Wonders of the World. Created in the first millennium B.C. for the homesick Persian wife of Nebuchadnezzar II (reigned 605-562 B.C.), they were a series of terraces supported on stone vaulting. The stonework was strong enough to support soil for large trees to grow successfully. A sophisticated irrigation system, drawing water from deep wells, kept the gardens verdant. In Ancient Egypt, gardens were portrayed in frescoes and carvings on tombs and temples, while in Israel King Solomon (c. 974-937 B.C.) wrote of pleasant gardens full of spice- and fruit-bearing plants and trees. The quest for spices led to the first recorded planthunting mission, despatched to "the land of Punt" on behalf of Queen Hatshepsut of Egypt in 1495 B.C. The expedition was led by Prince Nehasi, and sailed up the Nile before crossing to the Red Sea and south to Somalia, where trees of frankincense (*Boswellia*) and myrrh (*Commiphora*) were collected. They were carefully packed in baskets and taken back to Thebes to be planted in the garden of the Temple of Amun.

The gardens of ancient Egypt are said to be the foundation of western horticultural tradition, since they were a direct influence on the gardens of Greece and Rome. However, this is also true of the *paradeisoi* of Persia. The campaigns of Alexander the Great (356-323 B.C.) brought Greeks into contact with Eastern horticulture, and parks began to appear in Athenian Greece. In Rome, the landscaping tradition blended with the smaller, more practical *hortus*—a garden originally for the cultivation of edible produce, but soon adapted to include ornamental plants and features such as pools. This was gardening on a scale recognizable today, where intimate, private spaces were created for domestic pleasure.

The collapse of the Roman Empire in the 5th century A.D. left European horticulture in the care of a few monasteries, from where it was to emerge in the Middle Ages.

Elsewhere, other civilizations were developing their own gardening traditions. In western Asia and parts of the southern Mediterranean, Islamic gardens flourished, and were later a strong influence on resurgent

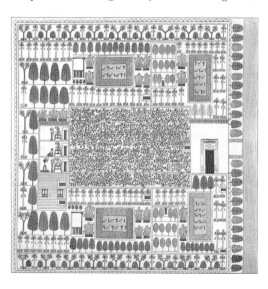

ORNAMENT AND USE *An Egyptian garden c. 1,400 B.C.
combines useful trees and vines with ornamental pools;
horticulture was already perceived as an ornamental art.*

Plant records

The Royal Botanic Gardens at Kew, near London is one of the most important reference collections of plants in the world, holding over 5 million specimens. Kew began life as a royal park, in which a small botanical collection was started in 1759, but its position in the forefront of botany and horticulture was established under the guidance of Sir Joseph Banks (1743-1820) from 1771 to his death. In his youth Banks had traveled widely, and understood the benefits of having field collectors to acquire new material. The next director to make a significant impact to Kew's reputation was Sir William Jackson Hooker (1785-1865), whose personal collection of dried plants was the foundation of Kew's Herbarium. Together with his son Joseph Dalton Hooker (1817-1911) who suceeded him, he extended the gardens to their present size and added the elegant Palm House in 1844-48, and Temperate House in 1860 providing essential glasshouse space for the tropical plants. More greenhouses have been added during this century, including the Alpine House and Princess of Wales Conservatory, both featuring installations to create optimal growing conditions for the plants. Another branch has been established at Wakehurst Place in Sussex, where the conditions permit a different range of plants to be grown and is the site of the plant physiology laboratory, and the seed bank, where germplasm of the world's flora is stored. Increasingly, Kew sees itself as a conservation resource, and its scientists are now as likely to be involved with conservation research as traditional taxonomy.

Sir Joseph Hooker

The Palm House, Kew Gardens

AUTHOR'S OWN
The planthunting tradition lives on: the author himself discovered <u>Impatiens kilimanjari</u> subs. <u>pocsii</u> (above) in 1990, and it was named only in 1997.

horticulture in Europe. Oriental gardening has pursued its own serene course for over 2,000 years, first in China and later in Japan. Here, form, texture, and symbolism were the most important aspects of the garden. Further afield, the Aztecs of Mexico and the Polynesians of the Pacific Islands practiced styles of horticulture which have since been almost totally lost.

The herb connection
One of the legacies of Rome and Greece to survive the Dark Ages was the rudiments of herbalism, the use of plants for medicinal purposes. The Romans had had extensive knowledge of this, and their armies had spread useful herbs throughout the empire. Classical naturalists, such as Aristotle (384-322 B.C.), Theophrastus (c. 370-286 B.C.), and Pliny the Elder (A.D. 23-79), had cataloged plants in their works, often recording their virtues as well as their names; they were among the first botanists.

The most important record of the medicinal uses of plants in ancient times is the illustrated *De Materia Medica* of Dioscorides, a surgeon in the Roman army of the 1st century A.D., but in medieval times, plant illustration became increasingly stylized (see p. 206). With the rebirth of learning—the Renaissance—from the late 14th century, updated herbals stimulated a widespread interest in plants. With the greater stability and prosperity of northern Europe, as

well as improved trade links, gardeners had access to a much wider range of plants than at any time in the past. Botanical gardens were founded across Europe to cultivate medicinal plants for use as teaching aids in the new universities. Species new to cultivation flooded in from the Orient and the New World.

PLANTHUNTING: A GLOBAL NETWORK
In 1597 the great English herbalist John Gerard (1545-1612) produced his *Herball*, which still attempted to provide medicinal information for every species, but also commented on their decorative merits. By 1629, John Parkinson (1567-1650) had relegated the "virtues" to a minor paragraph. His book, *Paradisi in Sole Paradisus Terrestris* (the title is a dreadful pun on his own name "The Park in the Sun, the Earthly Paradise"), describes 1,000 species of plant cultivated in England at that time. It tells us, for example, of the routes by which plants came into cultivation, and of the network of gardening friends in England and on the continent of Europe who corresponded with each other and exchanged plants, including the Tradescants, father and son, Peter Collinson, John Bartram in America, the Robins in France, and Carolus Clusius in Holland.

Such networks are still one of the most important ways in which plants are spread from garden to garden. I swap seeds and plants with gardeners around the world, and one is a living link with those Jacobean gardeners. The Plymouth strawberry is a curious form of the wild strawberry (*Fragaria vesca*) in which the floral parts are all replaced by tiny green leaflike organs. It was discovered in a garden in Plymouth, England, by John Tradescant the Elder, and described by Parkinson in 1629. In succeeding centuries it nearly died out several times, but someone always rescued it from extinction. A friend gave it to me and I have since passed it on to others. Fortunately, it is now also commercially available, making its future more assured.

Intrepid planthunters
The Swedish naturalist Carolus Linnaeus invented his binomial system (see p.219) at a time when only a small fraction of the world's flora was known. He believed, as did many others, that the tropics must have a rather sparse flora because he kept seeing the same plants in many different places. In fact, this was because they were common weeds. Within a few

years, however, European powers expanded into the hearts of the continents and found an incredible diversity. Nurseries, as well as botanical gardens, sent out explorers—planthunters—to bring back the choicest and rarest of plants, often at great risk to themselves.

Global village

Planthunting has become ever easier with time—air travel means that anyone can go to remote parts of the world—but it is not necessarily safer. While hunting on Mount Kilimanjaro, Tanzania, I often had to dodge elephants, and on one terrifying occasion I was chased into a tree by a charging lion. Today, however, planthunters cannot expect free access to botanical treasures; many plants are protected by law and collection permits are required in most countries. Despite restrictions, new plants continue to reach cultivation, enriching our gardens in the process.

Conservation of wild places and wildlife is increasingly important as the human population of the world rises inexorably. In the past, gardeners have often been guilty of excessive collection from the wild, but now habitat loss from urban expansion and improved agricultural techniques is the greatest threat. Instead, gardens can serve as reservoirs of plants, whose wild populations are threatened with extinction, from which reintroductions can be made in the future.

Patient plant breeders

Garden plants may be far removed in appearance from their wild ancestors, often as a

NATURAL SELECTION
The occasional white flower among a field of snakeshead fritillary (below) could provide the basis for a plantbreeder to select a white variation.

Life of a planthunter

David Douglas (1798-1834), one of the greatest planthunters of all time, was employed by the Horticultural Society of London (now the Royal Horticultural Society) to collect seeds in western North America. Between 1824-7 he collected the seeds of many coniferous trees, including the Douglas fir (*Pseudotsuga menziesii*) and the Sitka spruce (*Picea sitchensis*), as well as the flowering currant (*Ribes sanguineum*) and California poppy (*Eschscholzia californica*). In the process, however, he had to stave off attacks by Indians and nearly starved on several occasions. A second visit was equally profitable botanically, but in 1833 his canoe overturned in some rapids, all his specimens and equipment were lost, and Douglas himself was washed up, half-drowned. He re-collected the precious specimens and set out for England, calling at Hawaii on the way. There he continued to collect specimens until, walking through the forest alone, he fell into a pit-trap dug to catch the wild cattle that roamed the forests. Douglas' mangled body was found later—a bull was also in the pit.

result of a rigorous process of selection and breeding in cultivation. All wild plants vary slightly between individuals in every possible character, visible or otherwise—size, color, hardiness—the list is infinite. Such variation is the starting point for evolution, which, as Charles Darwin (1809-82) described in *The Origin of Species* (1859), occurs by a process of natural selection. Individuals with favorable characteristics are more likely to survive and reproduce than those in which a similar characteristic is less favorable, thus enhancing the proportion of those characteristics in the population. Eventually, in this way, sufficient differences may accumulate to make that population distinct, and thus recognizable as a taxonomic unit in its own right.

Gardeners use a similar process of selection to achieve the characteristics they want to see in their plants. However, if they mate two suitable parents, they can hasten the process. The progeny of the cross can be selected rapidly and the undesirables discarded (see p. 102).

ONLY THE ICE CAPS AND THE VERY
DRIET OF DESERT DUNES HAVE
ABSOLUTELY NO PLANTS—EVERYWHERE ELSE IS PLANT HUNTING
TERRITORY.

COLLECTORS' PARADISE

In 1495 B.C. the Egyptian Queen Hatshepsut sent the first recorded planthunting mission to the Land of Punt in eastern Africa on a quest for incense-bearing plants. Since then, botanists have cataloged some 250,000 species of flowering plants and ferns from almost every habitat on Earth, and some 50,000 of these are grown in gardens worldwide. In addition, there are innumerable garden-raised cultivars.

TAPPING EUROPEAN RESOURCES
Until the 1500s, the gardens of Europe contained very few "exotic" plants; most cultivated flowers were selected from the nearby countryside. Unfortunately, the Ice Ages had left northern Europe rather depauperate in plants, so the choice was relatively limited. Meadows, marshes, and lakes provided many beautiful plants, but gardeners hankered after the exotic. In part, their desire was satisfied by the selection of variants of wild plants, such as a white form or a double, which arise as spontaneous mutations, but new plants were always sought from abroad.

Not far away were the great flower-growing areas of Europe—the Alps and other mountains, and the Mediterranean basin. Montane plants, whether true alpines, or larger species from the subalpine hay meadows, had the advantage of being hardy in northern gardens. Although often similar to the wildlings, their introduction enlarged the possibilities available to the keen gardeners of northern Europe. From the Mediterranean, with its long, hot summers, came a tremendous selection of bulbous plants with subterranean storage organs, which permitted them to survive the dry season unscathed. Annuals survive by growing during the wet winter, flowering in the spring, and scattering their seed before the summer drought. Moister conditions further north allow them to be sown in spring to provide a display in summer. Until Classical times, much of the Mediterranean basin was covered in forests of

drought-tolerant trees. Many have become important garden evergreens—the stately columns of the cypress (*Cupressus sempervirens*) are seen as the epitome of Mediterranean gardening. Clearance has since converted forest to scrub—the garrigue or *maquis* vegetation whose aromatic shrubs perfume the warm air. Most of these species have toughened, often rather narrow leaves, adaptive features that reduce water loss in summer. Their fragrance is also adaptive, being a deterrent to browsing animals such as goats, but is a feature that has won them the admiration of gardeners. In addition to ornamental plants, many culinary herbs such as rosemary and thyme are also derived from the Mediterranean shrublands.

Despite this increased base of European natives, planthunters set out on expeditions to every corner of the world in search of plants to enrich the gardens of Europe, and later the rest of the world.

OUT OF AFRICA
The southern continents—Africa, Australia, and South America—have contributed relatively few plants to the gardens of the northern hemisphere. In each case, only a small proportion of their landmass is in temperate latitudes, and although there are mountains which support temperate plants, their contribution to northern garden flora has consequently been slight. A conspicuous exception is South Africa, where a diverse climate encouraged the development of fabulous biodiversity. Many of its species have been cherished by gardeners in other lands for over 300 years; others are only just appearing in cultivation.

The species-rich *fynbos*, or shrubland, is found only at the Cape; it includes many species of proteas and heaths, as well as bulbs and annuals. The spring displays of annuals in Namaqualand (Western Cape) are a famous tourist attraction, and many species have

MEDITERRANEAN NATIVE
Anemone coronaria painted by Pierre-Joseph Redouté grows wild in many parts of the Mediterranean region.

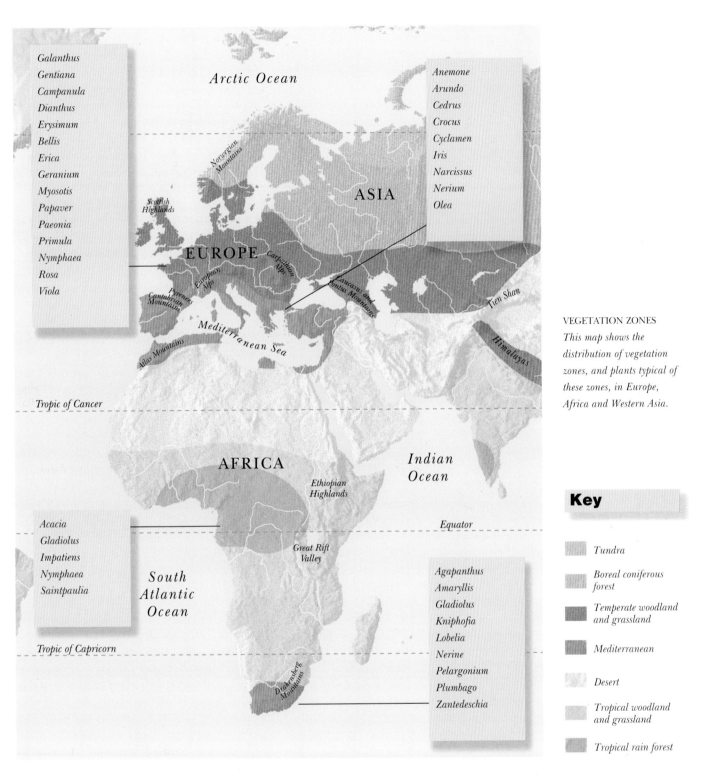

Galanthus
Gentiana
Campanula
Dianthus
Erysimum
Bellis
Erica
Geranium
Myosotis
Papaver
Paeonia
Primula
Nymphaea
Rosa
Viola

Arctic Ocean

Norwegian Mountains

Scottish Highlands

EUROPE

ASIA

Carpathian Alps

European Alps

Pyrenees
Cantabrian Mountains

Caucasus and Pontus Mountains

Tien Shan

Mediterranean Sea

Atlas Mountains

Himalayas

Anemone
Arundo
Cedrus
Crocus
Cyclamen
Iris
Narcissus
Nerium
Olea

Tropic of Cancer

AFRICA

Ethiopian Highlands

Indian Ocean

Equator

Acacia
Gladiolus
Impatiens
Nymphaea
Saintpaulia

South Atlantic Ocean

Great Rift Valley

Tropic of Capricorn

Drakensberg Mountains

Agapanthus
Amaryllis
Gladiolus
Kniphofia
Lobelia
Nerine
Pelargonium
Plumbago
Zantedeschia

VEGETATION ZONES
This map shows the distribution of vegetation zones, and plants typical of these zones, in Europe, Africa and Western Asia.

Key

Tundra

Boreal coniferous forest

Temperate woodland and grassland

Mediterranean

Desert

Tropical woodland and grassland

Tropical rain forest

become worldwide bedding plants. Within recent years, as patio planting has become fashionable, several new introductions from South Africa have become established, such as the many species and cultivars of the pink, trailing plant *Diascia* and the trailing *Sutera cordata* 'Snowflake' (also known as Bacopa).

The mountains of the Cape and the Drakensberg range have contributed some of the stalwarts of the herbaceous border,

especially for warmer gardens, including the African lily (*Agapanthus*). Collectors of succulents find the semidesert conditions of the Karoo a rich source of succulents, such as *Aloe*, *Euphorbia*, and the living stones, *Lithops*, whose perfect camouflage hides them from the attentions of thirsty animals.

Few of South Africa's trees and shrubs have been widely planted elsewhere, although they are often well-suited to semi-arid conditions and

VEGETATION ZONES
*The map opposite shows
the distribution of
vegetation zones across
Asia and Australasia*

CHIINA ROSE
*A wild rose of western
China was brought into
cultivation some time
before the 10th century
A.D., and named* Rosa
chinensis. *Painting by
Alfred Parsons.*

their merits recognized within South Africa itself. The primitive cycads are well represented in South Africa by some 30 species of *Encephalartos*, which have stout trunks bearing palm-like crowns of spine-tipped leaves. They were major constituents of the world's vegetation during the Mesozoic period (248-65 million years ago) and may have been food for dinosaurs. Today, cycads are restricted to corners of the tropics and subtropics. They only occur in small numbers and are threatened by development and the collection of plants for sale to unscrupulous gardeners, who covet their architectural qualities.

Tender tropicals

Tropical Africa has many attractive flowers, but few are widely grown. Most are tender and accustomed to the constant length of the tropical day, so they respond poorly to the seasons of temperate latitudes. Exceptions include the busy lizzie (*Impatiens walleriana*). Madagascar has supplied horticulture with prickly plants such as *Pachypodium*, admired by lovers of the bizarre, and the flamboyant tree *Delonix regia*, which is now very rare in the wild.

European relatives

The Atlas mountains, although on the African continent, have a largely European flora similar to that elsewhere in the Mediterranean basin. Their most famous plant is the blue-needled cedar, *Cedrus atlantica* 'Glauca', now commonly planted in North America. Many unique species have evolved on the isolated islands off the West African coast—the Azores, Madeira, and the Canary Islands—including the giant echiums.

ASIA AND AUSTRALASIA

The great Asian landmass, stretching from Turkey to Kamschatka and from the Arctic Circle to Cape Comorin, encompasses every conceivable habitat and is one of the most important sources of plants for temperate gardens. The Arctic has low tundra vegetation with ephemeral flowers hugging the ground among dwarf willows and birches. A little further south, the boreal forest, dominated by coniferous trees, covers most of Siberia and continues into Alaska and Canada. Plant diversity here is low, but on the southern edges, such as in the Altai mountains in southern Siberia, a richer meadow and alpine flora occurs. *Viola altaica*, an ancestor of the garden pansy, originated from this unlikely area.

Bulb-rich central Asia

The mountains of Siberia merge into the arid ranges of Central Asia, where annual rainfall is low and hot desert conditions occur in the lowlands. These mountains are the home of many bulbous plants, including the early-flowering *Tulipa kaufmanniana*, which has given rise to many brightly colored dwarf cultivars. The bulb-rich terrain stretches from Turkey to Pakistan, but at present is rendered largely inaccessible to the planthunter by unstable or hostile politics.

Central Asia was one of the areas where the garden culture of the Islamic world developed— Timur Leng, or Tamerlane (1336-1405), alternated ruthless war with creative cultivation at his capital in Samarkand. This gardening heritage passed to Mogul India, reaching its apogee in the terrace garden of the Taj Mahal mausoleum at Agra.

Much of India is too hot to provide many garden plants, although the holy pipal (*Ficus religiosa*) and banyan trees (*F. benghalensis*) are sometimes planted elsewhere in the tropics. The northern boundary of the Indian subcontinent is the Himalaya, source of many garden plants, including rhododendrons, primulas, and the blue poppies (*Meconopsis*).

The mountains then continue into China, and here, where China, Myanmar, and India meet, is some of the richest planthunting territory in the world. Planthunters such as Reginald Farrer, Frank Kingdon-Ward, and George Forrest established their reputations, stopped by nothing short of death. The botanical riches of the Sino-Himalayan mountain continue to be tapped even today, with recent introductions like the beautiful blue *Corydalis flexuosa* (Papaveraceae) from Sichuan. Although it was collected by Pere Armand David in 1896, it did not reach cultivation until the 1980s. First introduced to the U.S. in 1986 and to Britain in 1989, it proved to be so vigorous that, within a few years, hundreds of thousands of plants had been sold to amateur gardeners in both countries.

Riches of the Orient

China has contributed an enormous number of plants to the world's gardens, from the ancient conifers *Metasequoia* and *Ginkgo* to shrubs such as *Philadelphus*, *Skimmia*, and *Weigela*, as well as thousands of herbaceous species. Less affected by glaciation than northern Europe, the plant diversity of China is astonishing; the isolated

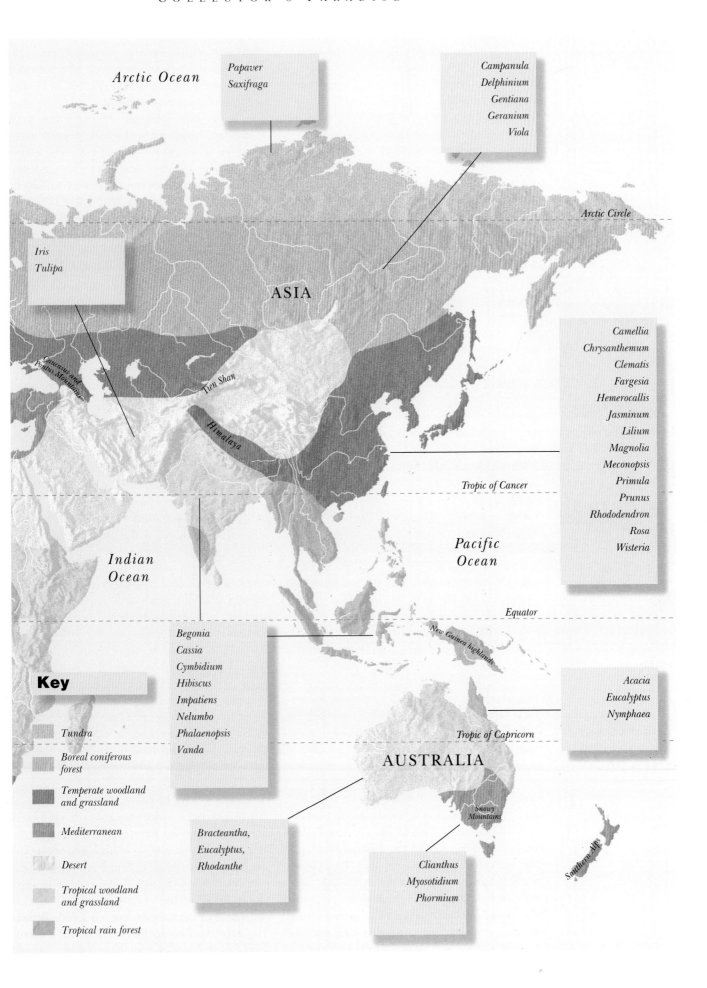

Arctic Ocean

Papaver
Saxifraga

Campanula
Delphinium
Gentiana
Geranium
Viola

Arctic Circle

Iris
Tulipa

ASIA

Caucasus and Pontus Mountains

Tien Shan

Himalaya

Camellia
Chrysanthemum
Clematis
Fargesia
Hemerocallis
Jasminum
Lilium
Magnolia
Meconopsis
Primula
Prunus
Rhododendron
Rosa
Wisteria

Tropic of Cancer

*Indian
Ocean*

*Pacific
Ocean*

Equator

New Guinea highlands

Begonia
Cassia
Cymbidium
Hibiscus
Impatiens
Nelumbo
Phalaenopsis
Vanda

Acacia
Eucalyptus
Nymphaea

Key

Tropic of Capricorn

AUSTRALIA

Tundra

Boreal coniferous
forest

Temperate woodland
and grassland

Mediterranean

Desert

Tropical woodland
and grassland

Tropical rain forest

*Snowy
Mountains*

Southern Alps

Bracteantha,
Eucalyptus,
Rhodanthe

Clianthus
Myosotidium
Phormium

Mount Omei in Sichuan boasts 3,000 species of flowering plants. Much of China has a temperate climate, and in subtropical areas there are high mountains with a rich temperate flora. The gardeners of ancient China and Japan (which is also rich in plants) used very little of this floral heritage, preferring more subtle effects derived from artful use of landscape features, and the form and texture of foliage plants. The few plants that were developed are garden classics—among them the chrysanthemum, tree peony, and camellias—with perfection achieved long before they were exported to the West.

Rainforest to desert in southern Asia

Southern Asia is tropical and was mostly covered with forests. Sadly, these have been ravaged during this century and now cover a fraction of their former area. With the trees have gone the orchids and palms coveted by collectors. The rainforests extend into New Guinea and Queensland, where there is a transition to the *Eucalyptus*-dominated vegetation of arid Australia. Many species of gum trees (*Eucalyptus*) are cultivated for their decorative bark, handsome flowers, and drought-tolerance. Australia shares its wealth of *Acacia* species with Africa, but unlike the giraffe-repellent thorn trees of Africa, Australian wattles are usually less spiny. Few Australian plants are hardy in temperate gardens, but the Kangaroo Paws (*Anigozanthus*) is often grown in frost-free conditions for its curiously shaped, somberly colored flowers.

The climate in parts of southeastern Australia and Tasmania is more temperate and supports lusher forests. From here comes *Dicksonia antarctica*, a magnificent tree fern which can be grown unprotected in warmer parts of Europe and the U.S. New Zealand supplied the world with hebes and phormiums, invaluable for maritime plantings because of their resistance to salt spray.

THE AMERICAS

Like China, the North American landmass is an enormous area of diverse climate and habitat which has provided many garden plants. In spring, the woods of New England are carpeted with charming small flowers, which are also welcome in gardens. Like most woodland plants they grow and flower before the leaves appear on the trees, and then retire into dormancy for the summer and winter. The trees of New England and the eastern U.S. are famed for their fall color, often provided by species of maple (*Acer*), and are widely planted in Europe and elsewhere for this feature.

In midcontinent the prairies have been largely replaced by farmland, but these ancient grasslands supported many fine herbaceous plants, such as the blazing stars (*Liatris*) with their tight purple or white columns of flowers. Prairie grasses, such as the bluestem (*Schizachyrium scoparium*) with its foxy-red winter color, are becoming popular for naturalistic mixed plantings of grasses and herbaceous plants. Prairie plants are extremely hardy and suitable for areas with harsh, continental climates with cold winters and hot summers.

The Pacific northwest has a wealth of trees which thrive in high rainfall, including the coastal redwood (*Sequoia sempervirens*) and its inland counterpart the giant sequoia (*Sequoiadendron giganteum*). Sequoyah was a Native American chief renowned for his environmentalism. In addition, there are lilies and lupins, and the mountains have shooting stars (*Dodecatheon*) and penstemons among many other rock garden delights.

Further south, the more arid climate of southern California supports the chaparral. Its aromatic scrub, interspersed with bulbs and annuals, recalls the Mediterranean garrigue or the South African fynbos. *Ceanothus*, *Fremontodendron* and the tree poppy, *Romneya*, grow here; in cultivation they complement their Mediterranean counterparts. Similar vegetation extends into Mexico, where there are also many species of shrubby *Salvia*.

Desert and scrub

The chaparral intergrades with desert vegetation, characterized in movies by the branched columns of the saguaro cactus (*Carnegiea gigantea*). The Cactaceae is an almost entirely American family—just one species occurs in Africa. Cacti have perfected their adaptation to a dry environment by evolving a water-conserving physiology, as well as spines which prevent browsing animals reaching their moisture reserves. The rosette-forming yuccas and century plants (*Agave*) also originate in the arid regions of the New World.

Southern diversity

The rainforests of South America, which include the most species-rich areas on earth, are the home of many orchids, bromeliads, and

AMERICAN INTRODUCTION *"Shooting stars," also known as the "American cowslip" —Dodecatheon meadia was collected on the prairies of Central America.*

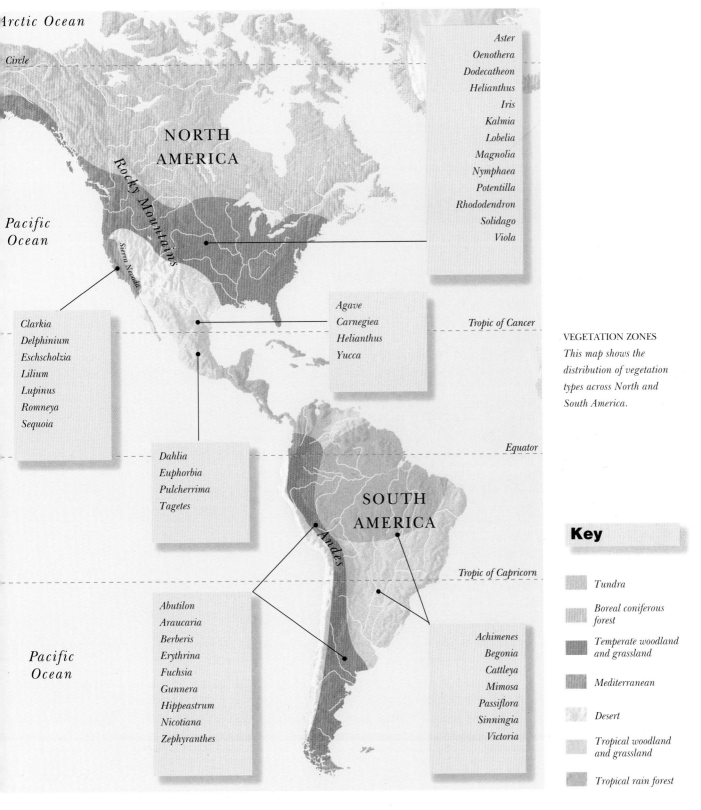

Arctic Ocean

Circle

NORTH AMERICA

Rocky Mountains

Pacific Ocean

Sierra Nevada

Aster
Oenothera
Dodecatheon
Helianthus
Iris
Kalmia
Lobelia
Magnolia
Nymphaea
Potentilla
Rhododendron
Solidago
Viola

Clarkia
Delphinium
Eschscholzia
Lilium
Lupinus
Romneya
Sequoia

Agave
Carnegiea
Helianthus
Yucca

Tropic of Cancer

Dahlia
Euphorbia
Pulcherrima
Tagetes

Equator

SOUTH AMERICA

Andes

Tropic of Capricorn

Abutilon
Araucaria
Berberis
Erythrina
Fuchsia
Gunnera
Hippeastrum
Nicotiana
Zephyranthes

Pacific Ocean

Achimenes
Begonia
Cattleya
Mimosa
Passiflora
Sinningia
Victoria

VEGETATION ZONES
This map shows the distribution of vegetation types across North and South America.

Key

Tundra

Boreal coniferous forest

Temperate woodland and grassland

Mediterranean

Desert

Tropical woodland and grassland

Tropical rain forest

houseplants such as *Philodendron* and *Monstera*. Long-tipped, leathery leaves are typical of many rainforest plants as they permit water to flow rapidly off the leaf's surface. In the Andes, tropical forest gives way to wet temperate forest where many species of *Begonia* and *Fuchsia* occur; in cultivation they flourish in cool, humid, but frost-free environments. Farther south are forests of bamboo, among which occur many flowering shrubs and herbaceous plants suitable for mild areas of the north temperate region. To the east are the grasslands of the pampas, where giant grasses eventually give way to shorter vegetation in Patagonia.

A FAMILY OF GOOD TRAVELERS,
FROM SIMPLE SNOWDROPS AND
DAFFODILS OF TEMPERATE AND MEDITERRANEAN REGIONS TO FLAMBOYANT
NERINE AND AMARYLLIS FROM WARMER CLIMATES.

THE NARCISSUS FAMILY
(AMARYLLIDACEAE)

Narcissus 'King Alfred'

Because they are bulbs (which have a natural period of dormancy after the flowers and leaves die back) the Amaryllidaceae are easy to transport, and this has led to their cultivation throughout the world.

The family name is derived from the ancient Greek word *amarysso*, meaning "to sparkle or twinkle"—an apt label, for many Amaryllidaceae flowers have an iridescent sheen. The leaves are strap-shaped, and many of the blooms are lily-like—some are known familiarly as lilies. An inferior ovary (situated

Hippeastrum psittacinum *Several, beautifully striped, small-* flowered wild species occur in the Andes. The larger cultivated versions look spectacular in cultivation in warm countries, and as indoor plants elsewhere, where they need a fertile but well-drained potting medium, and doses of liquid fertilizer when in growth. Although the hybrids are often sold as dry bulbs, it is advisable not to disturb established plants as they take some time to recover from root damage.

Zephyranthes candida grows in marshy places in its native South America, although in temperate gardens it needs drier, warm conditions to flourish. The flowers appear in fall, while the narrow, rushlike leaves remain green throughout much of the year. Other species have yellow, pink, or red flowers and grow as far north as the southern U.S., where they are also known as rain lilies from their sudden appearance after rains.

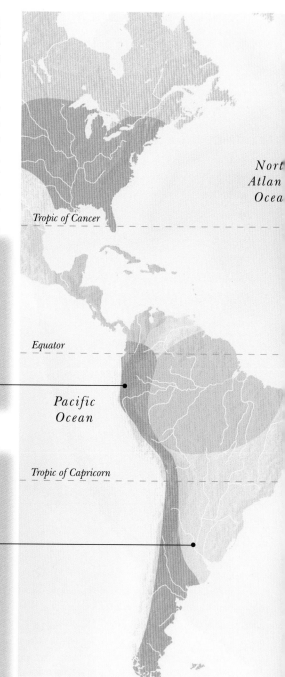

North Atlantic Ocean

Tropic of Cancer

Equator

Pacific Ocean

Tropic of Capricorn

below the petals) distinguishes them from true lilies which have a superior ovary above the petal line.

NARCISSUS: REFLECTIONS ON A NAME

Narcissus species, in particular, have been widely cultivated for centuries in Western and Oriental gardens because they travel well, multiply rapidly, and flower with cheerful reliability in spring. They are characteristic plants of the Mediterranean countries, where they grow and bloom in the wet weather of early spring, and then die back during the dry summers. Most wild narcissus are found in the

Narcissus bulbocodium *is from the mountains of France and Spain, and flourishes in moist, acidic soil, where it may also naturalize from seed.*

THE NARCISSUS FAMILY—
diversity around the world

Mediterranean Sea

likes moist conditions in cultivation, but thrives in any fertile soil where it is not disturbed. The larger L. aestivum *comes into bloom as its spring relation fades, and is a robust plant for waterside locations.*

Leucojum vernum, *the spring snowflake, is a native of damp places throughout Europe. It*

Indian Ocean

Key

uth ntic ean

where bulbs were taken to Europe in the 1700s, and to America during the 1800s. Flowering is triggered by the first heavy rains of fall, when the inflorescences appear rapidly, several months before the leaves.

Amaryllis belladonna *is a native of the Cape Province of South Africa, from*

	Tundra
	Boreal coniferous forest
	Temperate woodland and grassland
	Mediterranean
	Desert
	Tropical woodland and grassland
	Tropical rain forest

Parental responsibility

Over this century, daffodil breeders have greatly extended the range of colors available by using the red-edged cup of *N. poeticus*, a native of southern Europe, which, with *N. pseudonarcissus* is the parent of most modern garden daffodils.

Narcissus pseudonarcissus (*top*) which grows in drifts in damp fields and woods throughout western Europe, has many subspecies or closely related species which differ in size, form, and coloration. The Tenby daffodil (*N. pseudonarcissus* ssp. *obvallaris*) is pure yellow throughout, and one of the most elegant small daffodils for planting in grass or on the rock garden. It is known only from a limited area around the town of Tenby in south Wales and although it is assumed to have been introduced there, a wild counterpart has never been found anywhere else. However the parent of one of the most famous golden trumpet daffodils, 'King Alfred' (*above right*) is *N. hispanicus* from Spain.

NARCISSUS TRIANDRUS (far right), a native of Mediterranean countries, contributes its natural elegance and pendulous flowers to many garden daffodils.

NARCISSUS TAZETTA (below) was carried by traders from the countries of the Mediterranean to China and Japan, and is now the most widely distributed of all Narcissus.

wild in western Europe, especially in Spain, and were gradually introduced to northern Europe from the 1500s, where some species, such as the wild daffodil (*Narcissus pseudonarcissus*) have naturalized in woodlands and fields. A host of hybrids has been developed with colors ranging from deep golden yellow to combinations of cream, yellow, orange, and even salmon pink.

The narcissus story reaches far back into ancient history, for perfectly preserved specimens have been found in funereal garlands in excavated ancient Egyptian tombs. The ancient Greeks also associated the flower with death; according to myth, the sweet scent of *N. poeticus* enticed Persephone into the Underworld, and became a traditional flower to lay on graves. The flower is most associated with the story of Narcissus, a beautiful youth who, as a punishment for rejecting the advances of the nymph, Echo, was condemned by the goddess Aphrodite to fall in love with his own reflection. As he gazed adoringly at his image in a pool, he either died of despair because he could never catch hold of the object of his desire—or, according to another version, fell into the water and drowned. But a flower crowned with gold grew to cast its reflection on the water, and narcissus are even now often associated with waterside locations.

The word "narcissus" goes back beyond the myth. It is actually derived from the ancient Greek word *narce*, meaning narcotic, a reference to the drowsiness reputedly induced by the strong scent of some species. Although, the English language differentiates between the single-flowered form with a distinct trumpet— the "daffodil"—and "narcissus," which is used colloquially for smaller varieties and those with several flowers to the stem, there is no botanical difference between the two. The English word "daffodil" is misleadingly a corruption of *Affodyl*, the unrelated ashphodel; the Germans call the flower *Osterglocke*, or Easter bell, but in most European languages, the generic name *Narcissus* refers generally to all.

Eastern delights
The narcissus familiar to the ancient Egyptians and Greeks was probably the white

flowered, yellow trumpeted *N. tazetta*. This was probably the plant described in the *Hymn to Demeter*, sometimes attributed to the Classical Greek poet Homer:

"... wondrously glittering, a noble sight for all, immortal gods or mortal men; from whose root a hundred heads spring forth..."

N. tazetta is the most widespread narcissus of all and, in a reverse of the normal pattern, the only European plant to travel eastwards to become a favorite in China and Japan. Found wild throughout the Mediterranean, and cultivated in gardens throughout southern Europe and the Near East since ancient times, it was taken east by early traders, and has been well known there since the 10th century A.D. The Chinese call this fragrant, multi-headed species the "Good Luck Flower," or "New Year Lily," and now it is often forced to flower in time for the Chinese New Year celebrations. Its cultivation in China is centered on Fukien province, where legend tells of a poor old widow who was saving her last half bowl of rice for her feckless son, when a hungry beggar appeared requesting food. Out of pity she gave him the rice; as he finished it he spat out a few grains and vanished. In the morning the old woman found that the rice grains had been transformed into a narcissus in full bloom, and promptly found fame and fortune by selling the plants. Since then, *N. tazetta* has represented prosperity and benevolence.

N. tazetta is not frost-hardy, and is seldom grown in northern gardens, but the related "paperwhite narcissus," *N. papyraceus*, with pure white, intensely fragrant flowers, and the egg-yolk yellow *N.* 'Grand Soleil d'Or' are often grown indoors for early blooms.

Double wild

Double daffodils are not as popular as the single trumpet forms, but some have a long history in cultivation. In the spring of 1620, a fine double daffodil bloomed in the London garden of Vincent Sion, a Fleming, who had planted the bulbs a few years previously, but could not remember their origin. His plant is still called 'Van Sion', and although it has never satisfactorily been traced beyond that original garden, it has become widely naturalized throughout Europe and North America. It is found in situations that seem totally wild, although it clearly was originally cultivated, for the flower is completely sterile and can spread only by multiplication of the bulbs. The flowers

are a deep yellow, but very variable in form; one bulb may produce in successive seasons quite dissimilar flowers, some neat and regular, some ragged and untidy.

The delicate *N. triandrus* is a latecomer to horticulture, and was not widely grown until the 1800s. It is known in the English-speaking world as "Angel's tears," which seems suited to the delicate, pendulous flowers. One spring in the 1880s, Peter Barr (1826-1909), an important early breeder of narcissus hybrids, was prospecting the Spanish hillsides for new species. He sent his guide, a local boy named Angel Gancedo, to gather some *N. triandrus* bulbs on a hillside near the town of Corunna. Angel returned with the bulbs, exhausted and in tears, and so Barr labeled the bag "Angel's tears." *N. triandrus* has since become an important parent of hybrid narcissus, contributing its pale color and nodding flowers, although it is not easily grown itself. The vigorous, butter-yellow hybrid 'Hawera', raised in the New Zealand town of Hawera in the 1930s, retains all the dainty charm of *N. triandrus* and has rapidly gained worldwide popularity. Transporting plants between hemispheres can cause difficulties as the growth cycle adjusts to different seasonality, but narcissus adjust within a few seasons.

NERINE AND AMARYLLIS

Sometime in the 1650s, a ship believed to have come from Japan was wrecked off the coast of Guernsey in the Channel Islands. Some bulbs were washed into the sand dunes, and lay there unnoticed until they bloomed, with a startling display of bright red flowers, spangled all over with gold. Botanists recognized the strap-shaped leaves and inferior ovary characteristic of the Narcissus family, and at first called it "the purple narcissus from Japan," but it became known more enduringly as the "Guernsey lily." The story of the plant's discovery is commemorated in the scientific name *Nerine sarniensis*: in Greek mythology Nerine is a sea nymph, the daughter of Nereus, the Homeric "Old Man of the Sea," and Sarnia is the Classical name for the Channel Islands.

The genus in fact did not come from Japan at all, but from South Africa. The ship had probably called at the Cape of Good Hope en route back to northern Europe, and a crew member acquired the bulbs. The true origins of the plant were not established until the Scottish botanist Francis Masson found it growing in very infertile soil on Table Mountain, Cape Town, over 150 years later. In cultivation, as on the Guernsey dunes, *N. sarniensis* thrives in very sandy soil. It grows during the winter in the

WHEN THE RAIN FALLS
The seasonal rainfall of South Africa determines the growth patterns of Amaryllidaceae both in the wild and in gardens. Species from winter rainfall areas are usually tender in northern gardens, as they try to grow during the fall, when there is a danger of frost. Plants from summer rainfall areas are dormant during the dry, winter season, and are less likely to suffer from frost.

Key

Summer rainfall

Rainfall throughout the year

Winter rainfall

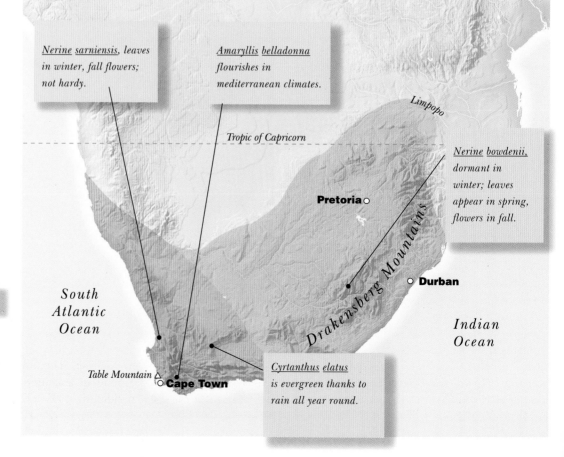

Nerine sarniensis, leaves in winter, fall flowers; not hardy.

Amaryllis belladonna flourishes in mediterranean climates.

Limpopo

Tropic of Capricorn

Nerine bowdenii, dormant in winter; leaves appear in spring, flowers in fall.

Pretoria ○

Drakensberg Mountains

○ Durban

South Atlantic Ocean

Indian Ocean

Table Mountain △
○ **Cape Town**

Cyrtanthus elatus is evergreen thanks to rain all year round.

THE "GUERNSEY LILY,"
was the name coined for
Nerine *sarniensis* *(above)*
after it was found among
the sand dunes in the
Channel Islands, though it
is a native of South Africa.
This cultivar, 'Major',
retains the iridescent sheen
of the species.

FLUORESCENT COLOR
In its native South Africa,
Nerine *bowdenii* *(left)*
forms swathes of bright
pink when it flowers in fall
after the summer rains. It
makes an excellent, long-
lasting cut flower.

north, and is too tender for outside cultivation, but can provide a magnificent greenhouse display in the fall. *N. bowdenii*, a much hardier mountain plant, is easier to grow. It was first found by Athelstan Bowden, a government surveyor, who sent some bulbs to his mother in Devon, England in 1889. After several years of growing them she sent some to the herbarium at Kew Gardens, London for identification, where they were found to be a new species and named after her son.

Naked ladies
In the South African fall, great tracts of land at the Cape are shot with startling pink by the massed blooms of *Amaryllis belladonna*. There are no leaves to lessen the impact for they do not appear until the flowers have finished—which has earned the plant the local nickname of "naked ladies." Flowering is most prolific after a bush fire which triggers the release of nutrients into the soil. The inflorescence spears from the ground in a matter of a few days, especially rapidly after rainfall, and the bracts open to reveal funnel-shaped, lily-like flowers which

POTTED GLORY
Hippeastrum hybrids, often much larger than their wild, South American parents, are popular houseplants and flower within a few weeks of being planted. With careful treatment, the bulbs survive and flower again the following year.

radiate from the top of the bare stem. The Swedish naturalist Carolus Linnaeus named the plant after Amaryllis, a shepherdess of Classical poetry and the Latin word *belladonna*, meaning "beautiful lady."

The name amaryllis is commonly given to the dramatic Hippeastrum hybrids which are grown as striking indoor flowering plants. These are derived from species which grow wild in southern and central America—such as *H. vittatum* from the Peruvian Andes. Although

they were originally included in the genus *Amaryllis*, they have since been transferred to their own genus which is only found in the Americas.

Several other amaryllids of South American origin have become popular garden plants in frost free areas of the world, such as *Zephyranthes*—"flower of the West Wind." The starry, white flowered *Z. candida* is the only species hardy enough for temperate regions, and even then only in milder areas, with a spot

in full sun. In the wild it is found in marshy ground near the rivers of Uruguay and Argentina, where, it is said, the sight of a million flowers in bloom inspired the Spanish conquistadores to name Argentina's great river Rio de la Plata—river of silver. The flowers appear between evergreen, rush-like leaves which can be damaged by frost, but in mild areas, *Z. candida* makes an attractive, fall-flowering companion to bloom at the foot of bare-stemmed nerines. Other rain lily species,

as they are called in North America, include the white-flowered *Z. atamasca* (a coastal plain native of the southeast), which is sometimes locally called the Easter lily.

SNOWDROPS FROM NORTH AND SOUTH
In the gardens of northern Europe, snowdrops are among the first flowers to appear in the early months of the year, but their English name seems to be derived not from climatic conditions, but from the German word

WHEN EUROPEANS SETTLED AT THE CAPE OF GOOD HOPE THEY CREATED A GARDEN; THIS WAS THE FOUNDATION OF A MODERN BOTANIC PARK WHICH BOTH CELEBRATES AND ENSURES THE CONTINUED SURVIVAL OF SOUTH AFRICA'S REMARKABLE INDIGENOUS FLORA.

Moraea loubseri

A GARDEN IN AFRICA

In 1652 the Dutch East India company established a permanent settlement at the Cape of Good Hope under the command of Jan van Riebeeck. His instructions included the establishment of a garden which would provide fresh vegetables to ships bound to and from the East Indies. By 1679, as newcomers to the Cape settled, ornamental plants were added, and native South African plants were sent to Europe from the garden. The Company's garden had become a staging post for a trade in plants that was to enrich two continents.

The Dutch East India Company's garden survives as the Cape Town Municipal Botanical Garden, but its role as an entrepot for plant distribution has passed to the South African National Botanical Institute at their principal garden, Kirstenbosch. The historic link between the two gardens is manifest in the survival, at Kirstenbosch, of a section of the hedge planted by van Riebeeck in 1660 as part of the defences around Cape Town. Kirstenbosch was founded in 1913 on a spectacular site covering 1,304

NATURAL FLORA
The Municipal Botanical Garden at Kirstenbosch preserves areas of natural Cape vegetation on the lower slopes of Table Mountain. The South African flora includes many succulents, such as Mesembryanthemum *(below right).*

1 Arboretum
2 Erica Garden
3 Pelargonium Koppie
4 Cycads Amphitheater
5 Indigenous Herb Garden
6 Vygie beds
7 Buchu garden
8 Camphor Avenue
9 Peninsula Garden
10 Vlei Garden
11 Van Riebeeck's Hedge
12 Protea Garden

SEASONAL CONTINUITY
Only about 7 percent of the whole area of Kirstenbosch is formally cultivated. The gardens are made up of ecological groupings planted with indigenous South African species. Though most of these are seasonal, careful selection has meant that there are displays of flowers throughout the year.

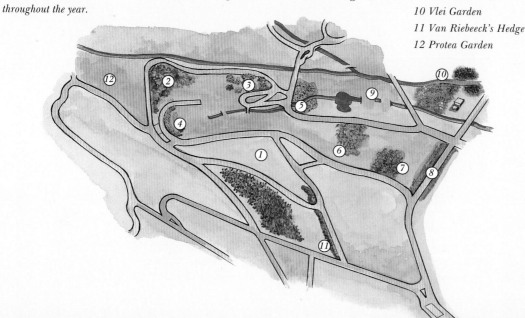

acres (528ha) of the slopes of Table Mountain. It specializes in the indigenous flora of South Africa, one of many botanical gardens around the world to emphasize their local botanical heritage. Most of the garden forms a reserve for native vegetation, such as the species-rich Cape shrubland, the fynbos, and coastal forest vegetation. The fynbos is unique to the Cape, but about one third of it has already been destroyed in its natural habitat. One of the glories of the shrubland, and of Kirstenbosch, are the stands of Silver Tree, *Leucadendron argenteum*, described by the Swedish botanist Carolus Linnaeus as "the most shining and splendid of plants." In spring, wild flowers provide brilliant sheets of color. Only 94 acres (38ha) of Kirstenbosch are formally cultivated, with indigenous species grouped together in appropriate conditions. One such section is the Cycad Amphitheater, where most South African species of this ancient group of plants are grown. In a new conservatory building, dryland and montane species are cultivated.

Conservation role

Expertise in cultivating indigenous plants has made Kirstenbosch an important center for conservation of the Cape flora. One small example highlights this vital role. Discovered in 1973, *Moraea loubseri* (Iridaceae) is a typical fynbos endemic, growing only on a small hill near Cape Town. When the hill was being quarried for railway ballast, specimen plants were taken to Kirstenbosch. Cultivation was succcessful, stock increased and was eventually distributed to other gardens. Wild stock still survives precariously, despite reports suggesting that it had been obliterated by blasting, but Kirstenbosch provides the insurance policy that guarantees the species' survival.

WINTER AWAKENING
Early spring snowdrops like Galanthus nivalis *(above) self-propagate once established. Many species release a fragrance when the air is warm. The bulbs dislike being dried out, so transplant the plants when in growth.*

IDENTITY MARK
The shape of the distinctive green mark on a snowdrop (right) is what enthusiasts use to identify the hundreds of different Galanthus *cultivars although in some the mark may be yellow, and in others missing!.*

Schneetropfen, a style of pendulous earring which was popular in the 1500s and 1600s. The scientific name, *Galanthus*, means "milk flower."

Although the 20 *Galanthus* species are very similar, with a distinctive white flower with interior green markings, there are over 600 cultivars. Many of these are distinguishable only by experts, but others show remarkable variations. One is the "yellow snowdrop," *G. nivalis* Sandersii Group, in which the interior marks are yellow. It grows wild in the woods of Northumberland in northern England, but although it is highly sought, it is not very strong and tends to die out unless administered with specialist care.

Soldiers in the Crimean War (1853-1856) were enchanted by sheets of snowdrops appearing on the ravaged battlefields in spring. These were *G. plicatus*, a larger plant than the common European *G. nivalis*. Many solidiers took bulbs home, establishing the species in British gardens. In Lord Clarina's garden in Straffan, Ireland the species hybridized with the local *G. nivalis*; among the progeny was one seedling of great beauty, bearing two flowers instead of one from a single bulb, which was eventually given the name 'Straffan'.

Another southern European species, now widely grown, is *G. elwesii*, from western Turkey. In the 1980s, this was collected from the wild at a terrifying rate of over 30 million per year to satisfy the horticultural demand. It was discovered in 1874 by Henry Elwes, an English landowner who botanized in between shooting big game and collecting butterflies.

There is a fall-flowering snowdrop too, which, like its warm climate relations—the nerines and amaryllis—blooms before the leaves develop. On the mountains of the southern Peloponnese in Greece, *G. reginae olgae* blooms from October, and is found throughout the lands bordering the Adriatic Sea, where it has been traditionally used for marriage wreaths.

The fall-flowering snowdrop was discovered by Greek botanist Theodoros Orphanides (1817-1886) in 1874, who did not tell anyone where he had found it. When he fell mentally ill, the plant's origins remained a mystery until it was rediscovered in 1896, and named after Queen Olga of Greece, grandmother of Prince Philip, Duke of Edinburgh. Unlike the hardy *G. nivalis*, it requires warm, well-drained conditions, and is often grown under protection in northern countries.

LEUCOJUM THROUGH THE SEASONS

Leucojum, or snowflakes, look like tall snowdrops, but a snowdrop has two whorls of segments of unequal length, while snowflake segments are of equal length. The spring-flowering *Leucojum vernum* has flowers like elegant miniature lampshades fringed with green or yellow. It grows wild in damp places throughout Europe, and, like its summer-flowering relative, *L. aestivum*, is often grown in bog gardens. *L. aestivum* is a much larger plant which may reach 3 feet (1m), and carries up to seven flowers on each stem. It grows in marshes from southern England to north Africa and east to Turkey. A selected clone with particularly large flowers is called 'Gravetye' after the garden in southern England of William Robinson (1838-1935), the "father" of the modern "English garden" style of horticulture. By advocating informality against the excessive rigidity of Victorian horticulture, Robinson and his friend Gertrude Jekyll (1843-1932) revolutionized gardening throughout the world.

The slender *L. autumnale*, from southwestern Europe and northern Africa, emerges with the first rains of fall, the fragile bells of crystalline white with a pinkish tinge appear before the leaves.

LIVELY AND BUSTLING PLANTS
WHICH BRING BRIGHT TROPICAL
COLORS TO BALCONIES, WINDOW BOXES, AND SUMMER BEDDING DISPLAYS.

BEGONIAS, BUSY LIZZIES, & OTHER BEDDING PLANTS

(BEGONIACEAE, BALSAMINACEAE, GESNERIACEAE, SOLANACEAE)

Saintpaulia cultivar

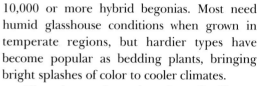
Begonias are named after Michel Begon (1638-1710), governor of the then French colony of Canada, and of course, a keen botanist. The genus *Begonia* includes more than 900 species found throughout the tropics, especially in South America. From these wild ancestors, gardeners have created an incredible 10,000 or more hybrid begonias. Most need humid glasshouse conditions when grown in temperate regions, but hardier types have become popular as bedding plants, bringing bright splashes of color to cooler climates.

The tuberous begonias, plants familiar to the temperate gardener's tub, hanging basket,

Achimenes *are found wild in the forests of Central America where species with orange, red, or purple flowers occur. Their common name of "hot water plant" comes* *from an old practice of watering the plants with warm water to encourage growth, but this is unnecessary in heated homes. Like* <u>Sinningia</u>, *they become dormant in winter, producing lots of small, grub-like tubers which provide an easy means of propagation.*

Sinningia speciosa *and its many cultivars are still widely known by their former scientific name of* <u>Gloxinia</u>. *Found wild in the tropical forests of Brazil, they appreciate warm, humid conditions and are often grown as house plants. The cultivars differ from the wild ancestor in their* *large, upward-facing flowers. As these bells easily fill with water, the plants should always be watered carefully. After flowering they die back to a dormant tuber, which should be kept dry until the following spring.*

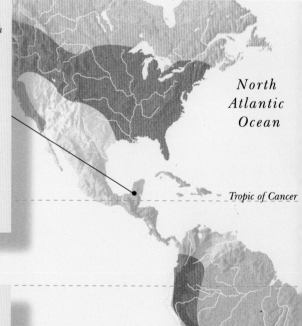

North Atlantic Ocean

Tropic of Cancer

Pacific Ocean

or summer bedding displays, form one of the main cultivated begonia groups; they are the favourite flowers of Kim Jong Il, current leader of communist North Korea. Defying all taxonomic conventions, he declared that the plants should be renamed "Kimjongillia," and despite his country's troubled economy, he ordered that vast tracts of land should be given over to begonia cultivation.

The ancestors of the tuberous begonias, a group of wild species from the cool forests of the Peruvian and Bolivian Andes, are found at altitudes of around 10,000 feet (3,000m). The cultivated descendants still prefer cool growing conditions like those of these mountain habitats.

Begonia boliviensis, a trailing species with bright scarlet flowers, was the first to reach Europe in 1847. Its speedy popularity ensured that others soon followed—*B. cinnabarina*, erect and orange-bloomed, *B. pearcei* with yellow flowers and a compact habit, and pink or white *B. rosaeflora*. As these originally distinct species came together in European nurseries, during the 1860s and 70s, hybrids soon appeared, with

a double-flowered cultivar by 1882. Systematically, breeders began to aim for many rounded flowers on a compact, robust plant. This was achieved when one auspicious individual bore flowers shaped like rosebuds.

Specialist selection

The parentage of hybrid begonias was soon so complex that they were all christened *B. x tuberhybrida*, followed by the name of their distinctive line. Blackmore and Langdon nurseries in Somerset, England, began breeding show begonias in 1901, and still produce some of the most outstanding large-flowered cultivars. However these and other lines developed by specialist nurseries need great skill and attention to perfect and reproduce. Commercial

Begonia masoniana, the "Iron Cross" begonia, earns its popular name from the resemblance of the dark marking on its leaves to the German military decoration for valor. It is a native of southeast Asia and one of many Asian begonias valued for its attractive foliage. The flowers are insignificant and are usually removed from specimen plants.

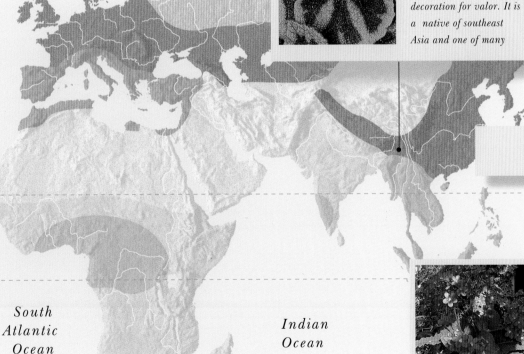

PLANTS FROM WARM PLACES

South Atlantic Ocean

Indian Ocean

Streptocarpus is a genus of about 120 species in Africa and Madagascar, where the plants usually grow in shallow soil in humid conditions. Many fine cultivars have been developed from South African species such as S. rexii, with flowers in shades of white, blue, purple, and red. An interesting feature is their ability to grow from leaf cuttings, which make it easy to grow a large number in a short space of time.

growers in France and Holland developed strains that could be grown from seed and sold as single colors. These include the 'Multiflora' types, also sold as corms for spring planting. More recently 'Non-stop' and 'Clip' have allowed amateurs to raise their own plants from seed each year. These plants are highly bred, yet retain evidence of their origins in habit and flower color. For example, 'Pendula' cultivars have the trailing habit of the wild *B. pendula*, while others inherit their stocky habit from the ancestral *B. pearcei*.

In contrast to the stout *B.* x *tuberhybrida* cultivars is charming *B. sutherlandii*, a slender, trailing species with soft orange flowers from southern Africa. Its name honors Dr. Peter Sutherland (1822-1900), government geologist and Surveyor-General of Natal in South Africa, who managed to integrate botanical activities into his official geological duties. This species grows wild on damp rocks, its small tuber enabling its survival during the dry season—and northern winters on the window sills of Europe and North America.

Versatility in flower and foliage

The small, tufted blooms of *B. semperflorens*, in white, pink, or red, brighten summer beds throughout Europe, North America, and other temperate regions of the world. These fibrous-rooted plants are yet another group of complex hybrids, derived from several South American species. They include *B. semperflorens*, a tall, white-flowering summer type, the shorter *B. schmidtiana* with occasional flushes of pink flowers, and *B. roezlii*, bright scarlet in winter and spring. Hybridization among them has produced a wide range of colors and a long flowering season.

Further selective breeding with *B. fuchsioides* produced the compact, small-leaved begonias of today. Although true perennials, they are usually raised annually from seed. Most are supplied as F1 hybrids (see p. 102), created to produce uniform plants. This work has been concentrated in Denmark, Holland, Germany, and North America.

Many begonias are grown as houseplants, appreciated not for their blooms but for their handsome foliage. The popular *B. rex* and its hybrids derive from a seedling in a potted orchid sent from India to England in the mid 1800s. *B. rex* and the close relations used for its hybrids are native to the Himalayan foothills of eastern India, where they grow at altitudes of

FOREVER FLOWERING
Begonia semperflorens (right), although strictly perennial, is usually grown from seed each year. By careful selection breeders have created cultivars with dark foliage and pale flowers, and variegated forms which come true from seed are now available.

700-3,000 feet (200-900m) in an area of very high annual rainfall. In cultivation they like high humidity, and moisture at their roots—as long as it is not stagnant—and they thrive best in good light but not in direct sunshine.

The *B. rex* group of rhizomatous begonias are valued for their infinitely variable, finely marked leaves which have zones and bands of red, dark green, pale green, white, or silver, in a variety of surface textures. The flowers are quite insignificant by comparison, and often removed.

In the 1950s, English amateur gardener Maurice Mason (1912-1993) collected a then unknown species in New Guinea. *B. masoniana* has attractive bright green foliage marked with a cross-shaped, chocolate-colored blotch—which gave rise to the popular name of "Iron Cross Begonia."

BUSY LIZZIES: IMPATIENT STARTERS

All members of the plant genus *Impatiens* have explosive seed capsules, with rapid-action fire that scatters the seeds several feet from the parent. This feature provides the scientific name, which simply means "impatient", as well as various common names, including "busy lizzie" and "touch-me-not." Other species, usually annuals with big, baggy flowers, are often known as "balsams."

Impatiens has some 850 species, almost as many as *Begonia*. A difference is that very few *Impatiens* are seen in cultivation. This is largely because their seeds have short life spans and seldom survive the journey from the wild to cultivation. The mountains of East Africa and Madagascar have many beautiful species with great horticultural potential, as do the Himalayas and other Asian ranges. Overcoming

COOL SPOTS
Hybrid tuberous begonias (left), derived from wild Andean species, enjoy cool conditions in tropical gardens, where they are best planted in semi shade.

PINK VARIATIONS
Busy lizzies (above) are derived from Impatiens walleriana *from the wet forests of East Africa, where the flowers are usually pink, but variants occur from which other colors have been selected.*

I. hawkinsii has given rise to the 'New Guinea' hybrids. These resemble *I. walleriana* but have a bushier habit and longer, darker leaves, frequently variegated with cream or red blotches. The flowers are also larger and more rounded than the *walleriana*-derived, common busy lizzies, but this distinction blurs further each year as breeders continue to adapt the common varieties.

Another large-flowered East African species, *I. sodenii*, forms bushes up to 5 feet (1.5 m) tall, with fleshy stems bearing terminal whorls of leaves and pale pink flowers up to 3 inches (8cm) across. It thrives in warm gardens worldwide and is known in Australia as the "Poor Man's Rhododendron" from its shrubby appearance and easy propagation; cuttings root if simply placed in a jar of water. Wild *I. sodenii* grows in very exposed, rocky places, and so is tolerant of full sun, provided the soil retains some moisture. However, like all *Impatiens*, it is frost-tender and must be kept warm through the cold winters of temperate regions.

A serious yet attractive weed

The problems of importing new species of *Impatiens* are illustrated by the tall annual *I. glandulifera*. Introduced into Europe in 1839, it escaped cultivation and now thrives as the serious, if attractive, riverbank weed known as Himalayan balsam. It is also called "Policeman's helmet" from its bulbous, rounded flowers, which resemble the tall protective headgear of the British police constable or "bobby".

Impatiens glandulifera demonstrates the complicated flower morphology of many *Impatiens* species, in which a sepal becomes enlarged into a curved spur behind the flower. These spurs vary in shape and are thought to assist pollination by bees, butterflies, or birds. For example, the red and green flowers of *I. niamniamensis*, which grows in the forests of central Africa, resemble a multicolored parrot and are probably bird pollinated in the wild. The species is named after the Niam-Niam, a southern Sudan group of people from the 1800s with cannibalistic tendencies. It is best cultivated in a greenhouse, although it thrives outdoors in tropical climates.

SAINTPAULIA, OR AFRICAN VIOLETS

The Usambaras, a group of ancient mountains lying inland from the port of Tanga in northern Tanzania, are one of the most biologically unique and diverse parts of tropical

the seeds' physiological problems would allow gardeners to explore this potential.

"Busy lizzies" (or "bizzy-lizzies") are mostly derivatives of East African *Impatiens* species, chiefly *I. walleriana*, with brightly colored, flattened flowers. *I. walleriana*, the original name, was sometimes known in cultivation as *I. sultani*. This later name honored the Sultan of Zanzibar, from whose territory it was introduced to Europe in 1896, but is now incorrect.

The original wild busy lizzies were rather straggly plants, with sparse foliage and few, typically harsh-pink flowers. In the hands of European plant breeders they have become today's compact, multiflowered, multicolored plants. Their cool, moist natural habitat has suited them to the climates of northern Europe and parts of North America, where they smother themselves in summer flowers of bright red, white, orange, even bluish-pink. Double-flowered cultivars are known, but these are usually propagated from cuttings, since double flowers seldom set fertile seed. The very variable

Africa. In 1892 the Governor of the area's Usambara District, then in German East Africa, was botanically-inclined Baron Adalbert Emil Walter Redliffe le Tanneux von St. Paul-Illaire—known as "Baron Walter" for short. While the Baron was exploring locally, a small fleshy plant with bluish-purple flowers caught his attention. He gathered seeds and specimens from the two colonies he found and sent them to his father, Baron von St. Paul-Illaire the Elder. The father was also a keen amateur botanist, and forwarded them to his friend Hermann Wendland (1825-1903), the director of Berlin's Royal Botanic Garden.

Within a year Wendland grew plants from the seeds and described them as a new genus, *Saintpaulia*, after the family who had discovered them. The species was named *Saintpaulia ionantha* from the Greek *ion* (violet), and *anthos* (flower), since the five-petaled flower resembled *Viola*. However, *Saintpaulia* is a member of the Gesneriaceae family, which is quite different from Violaceae, usually having rosettes of fleshy leaves and the corolla lobes of the flower united at the base (violas have separate petals).

African plants in America

In about 1925, Armacost and Royston, a Los-Angeles-based nursery, recognized the possibilities of *Saintpaulia* as a houseplant and acquired seed from Europe. They grew a thousand seedlings, and discarded all but the best hundred which were grown on for further observation. After further extensive trials the finest ten plants were selected and named. Almost all of the many thousands of African violet cultivars available today, which form a global horticultural industry, have been bred from that final ten. One of them, 'Blue Boy', became a particularly important parent in its own right. First it gave seedlings with red flowers; then in 1939 it produced a sport with double blue flowers; in 1940 a single pink seedling was named 'Pink Beauty': and in 1941

COLOR ATTRACTION
Impatiens niamniamensis (above) grows in the shade of African forests where it is pollinated by sunbirds probing for nectar in the curved spur. It grows outdoors in the tropics, but needs frost protection in temperate regions.

a seedling arose in which the paleness of the leaf stalk spread into the blade, resulting in a two-tone leaf.

These mutations were used in further breeding work which has led to the astonishing range of "African violets" or *Saintpaulia* seen today on windowsills around the world. Flower size has increased from just over an inch (3cm) in the wild species *S. ionantha*, to 3 inches (8cm) in some of the modern cultivars. Similarly, leaf shape now includes contorted, wavy-edged, and variegated forms. During the 1960s plant breeders interbred other wild species for more characteristics. The tiny species *S. pusilla* and *S. shumensis* helped to create miniature African violets when crossed with the familiar *S. ionantha* cultivars, while hybrids with *S. grotei* created trailing plants.

Common but threatened
African violets are still known as *Usambara veilchen* (violet) in Germany, recalling the

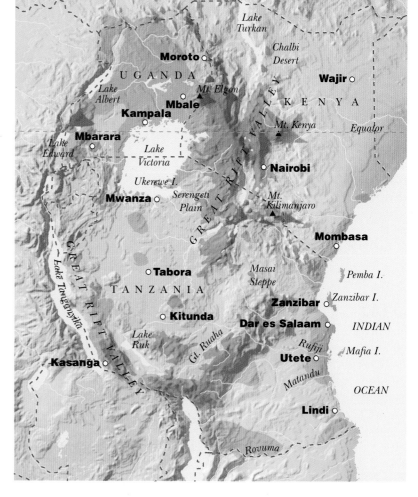

THIRST IN A DRY LAND
The mountains of East Africa (left) provide islands of dampness in the surrounding dry bush, where Impatiens *and* Saintpaulia *species can flourish.*

Rainfall over 40 in (100 cm) p.a.

Distibution of *Saintpaulia*

Land over 6,000 feet (1,500 m)— and distribution of Impatiens

cloudforest mountains of their home. Sadly the rich Usambara forests have been decimated for timber and tea plantations and the cosy humid understory favored by the wild species have largely disappeared. Despite its incredible familiarity and abundance in cultivation, *S. ionantha* and the other 20 species of its genus are threatened with extinction in the wild.

POPULAR PETUNIAS

Petunias, members of the Nightshade family, Solanaceae, are named from the Brazilian word *petun* (tobacco). However in that country, this name is applied not to petunias, but to the closely-related *Nicotiana* (see p. 216).

Petunias are primarily bedding plants with large, funnel-shaped flowers, but the wild species of the genus *Petunia*, from which the cultivated types are derived, are all natives of open places in South America, and rarely seen in cultivation. The first hybrids, between purple *P. integrifolia* and white *P. axillaris*, were produced in Europe from the 1830s onward and developed into a multitude of cultivars. Most petunias sold today are F1 hybrids, such as the 'Carpet' series. These guarantee compact, floriferous plants, being the end product of years of breeding to enhance characteristics such as weather tolerance.

A continuing supply

During the past decade, trailing petunias, especially the 'Surfinia' group, have come to the fore as plants for hanging baskets and window boxes. They produce hundreds of flowers during their long season, but unlike the compact bedding cultivars, these trailers do not breed true from seed and must be propagated annually from cuttings.

Even more recent petunias are the 'Million Bells' or 'Carillon' types. These form bushier plants bearing many smaller flowers in shades of pink or purple, with a yellow throat. They are so new that they have not yet reached the attention of most gardeners, but they look set to maintain the popularity of these bright tropical plants.

Campanula rotundifolia

𝒢ENTIANS AND 𝐵ELLFLOWERS

(GENTIANACEAE AND CAMPANULACEAE)

Standard issue blue gentians are found throughout the mountain ranges of the northern hemisphere. *Gentiana acaulis* from the Alps is the familiar trumpet gentian, and the adopted emblem of the Alpine Garden Society, whose worldwide membership is dedicated to the cultivation of alpine plants. *G. acaulis* has been cultivated since the 1600s, although it is sometimes reluctant to flower well.

Good parents

The rock gardening craze early this century led to the introduction of many other gentian species, especially from the Himalaya and China. One of these, from the barren uplands of Gansu Province, China, was the fall flowering *G. farreri*, described by its discoverer Reginald Farrer, as a "marvel of luminous loveliness."

Gentiana farreri has long trumpets of pale but intense sky blue, striped with black-purple and buff panels on the outside. The plants discovered by Farrer in September 1915 did not set seed before winter, so in desperation he took them back to England via the long Trans-Siberian railroad journey from Peking; none survived. Farrer became involved in First World War work, and forgot about the failure. Then, nearly a year later, he received a package from the Royal Botanic Garden, Edinburgh, Scotland, with a covering note asking him to identify the contents which had been grown from seed which Farrer had sent there some time before. Inside were perfect specimens of *G. farreri* in bloom. The seeds had been gathered in 1914, in a completely different locality, and misidentified. When crossed with other species such as *G. sino-ornata*, *G. farreri* became the parent of a race of fine garden plants, whose only drawback is their requirement for moist, acidic conditions.

An earlier introduction to European gardens was *G. septemfida*, a robust, sprawling

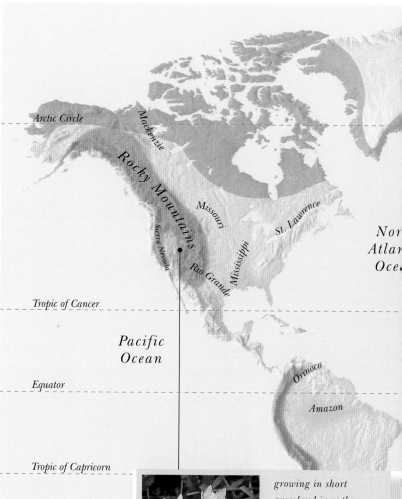

Key

Tundra

Alpine regions

Gentiana algida, *widely distributed in Siberia, northeast Asia, and the Rocky Mountains of Canada and the U.S., usually growing in short grassland in rather damp places. Ivory flowers, faintly dotted in Prussian blue, appear in late summer when most other alpine plants have finished flowering. In cultivation it needs a humus-rich soil and succeeds best where summer temperatures remain cool.*

Gentiana verna, *the spring gentian, grows in sheets in the alpine meadows of Europe, from Ireland to the* Caucasus Mountains. It emerges soon after the snow melts, and its piercingly blue flowers must attract pollinating insects before the grass grows up around them. In cultivation, the plant thrives in rich, moist, but well-drained soil.

Gentiana sino-ornata, *a fall-flowering trumpet gentian from western China, where it grows in alpine turf at altitudes up to 19,000 feet (5,800m). In cultivation it requires acid soil and a humid atmosphere. Many fine* hybrids have been raised from it, all requiring similar growing conditions.

ALPINE REGIONS OF THE WORLD—
the home of wild gentians

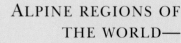

Norwegian Mountains

Scottish lands

Volga

Carpathian Alps

Danube

Caucasus and Pontus Mountains

Allai

Tien Shan

Euphrates

Himalayas

Indus

Ganges

Yangtze

Mekong

Mountains

Niger

Ethiopian Highlands

East African mountains

Congo

New Guinea highlands

Pacific Ocean

Zambezi

Limpopo

Drakensberg Mountains

Murray

Snowy Mountains

Southern Alps

Gentiana saxosa *grows near the sea. It is often shortlived and should be raised regularly from seed. Like many New Zealand plants, it has white flowers, presumably a response to the insect pollinators available.*

Gentiana lutea, *the statuesque yellow gentian which can reach 6 feet (2m) in height. The large foliage is* strongly veined and attractive in its own right, and the starry flowers are borne in whorls up the stem. It is a slow-growing perennial which needs rich soil and patience in cultivation.

species that is easy to grow in European gardens, and which produces a profusion of blue trumpets in late summer. It is so amenable that Farrer called it "a friend of man." *G. semptemfida* was discovered in the meadows of the Caucasus mountains by Count Apollon Apollosevitch Musin-Puschkin (*d.*1805), a former Russian ambassador to London, and the botanist Marschall von Bieberstein (1768-1826) These two men introduced many Caucasian plants to western Europe, including the tall *Campanula lactiflora*, which reaches 6 feet (2m) , and bears clouds of pale blue, upward-facing flowers during summer.

Medicinal origins

Although gentians are valued today as decorative plants, they were originally considered important medicinal plants. They are named after King Gentius of Illyria (modern day Yugoslavia), who ruled from 180 to 167 B.C. and is accredited with discovering their medicinal properties. However, an ingredient on a prescription found with a mummified ancient Egyptian which long predated King Gentius has also been ascribed to *Gentiana*. The medicinal extract was very bitter; it was used to flavor beer before hops, and is still used in some liqueurs. The species most often used medicinally was *G. lutea*, a statuesque perennial common in the subalpine meadows of Europe. This is not easy to grow and is seldom seen in cultivation.

Although most gentians are blue, several species have yellow flowers. In New Zealand, all gentians are white, including *G. saxosa*, a low, mat-forming species which bears a profusion of flowers, but is apt to die after a heavy flowering. In the Andes grow several species with red, tubular flowers.

CHINESE PARENTAGE
*Gentiana farreri (left) was
discovered on the
mountainsides of western
China by Reginald Farrer,
and became an important
parent of garden hybrids.
It needs acid soil.*

EURO TRUMPET
*The typical blue trumpet
gentian of European
mountains is Gentiana
acaulis (left). It often
grows in company with G.
verna, which flowers a
little later in the year, and
will tolerate some lime in
the soil.*

Reginald Farrer:
alpine garden wizard

Yorkshireman Reginald Farrer (1880-1920) was almost singlehandedly responsible for development of the art of rock gardening, and known to his contemporaries as the "rock garden wizard." He conveyed his enthusiasm in unparalleled purple prose— his book *The English Rock Garden*, published 1920, remains one of the standard works on alpine plants, and is valued as much for its flamboyant descriptions as for its information. He scorned the "dog's grave" and "almond pudding" styles of suburban rockeries, prescribing a completely new formula for rock garden design.

After forays in the European Alps, and a trip to Japan during which he became a Buddhist, Farrer set out for northwestern China in 1914 with fellow enthusiast William Purdom (1880-1921). Among many notable plants introduced by the expedition was the widely planted *Viburnum farreri*, a fragrant, winter-flowering shrub. In 1919 Farrer set off for Burma with Euan Cox, but the expedition was less productive as the area visited was too warm to provide many plants for European gardens. Cox returned home in early 1920, but Farrer persevered in the dripping conditions of the Burmese forest. He became ill, and died on October 17, 1920, alone except for his servants who later rescued his notes and paintings—but not the seed harvest!

CAMPANULAS: FAMILIAR FRIENDS

The 300 species of *Campanula* vary in stature from minuscule plants found in rocky crevices to leafy perennials reaching over 6 feet (2m) tall. All are immediately recognizable by the five-lobed bellflowers in shades of blue, purple, or white.

When the English missionary Charles New was exploring the luxuriant flora of Mount Kilimanjaro in East Africa in the 1870s, he wrote of his great relief in finding "such old friends as the dock and common stinging nettle." Similar feelings are aroused when one encounters another, and rather more attractive, "old friend," the harebell, *Campanula rotundifolia*, in unfamiliar surroundings. It varies considerably in color tone and growth, but is always instantly recognizable: an archetypal bellflower, with large, pale blue thimbles borne on wiry stems. A tuft of bellflowers in the gravel of a riverbed in the Altai mountains of Siberia turned out to be *C. rotundifolia*, as did straggly shoots in the mountains of California. Wherever it grows, *C. rotundifolia*, as with all alpine bellflowers, occurs in conditions of perfect drainage; in cultivation they need rock garden conditions to do well.

Although campanulas are characteristic plants of the mountain ranges of the northern hemisphere, half of all the species occur in Mediterranean countries, often growing in crevices exposed to full sun and forming rosettes that may grow for several years before flowering. Others are perennials from woodland or grassland habitats, while some grow as annuals or biennials in disturbed ground.

Ringing the changes

Campanula medium, although a native of southern Europe, is known throughout the English-speaking world as the Canterbury bell, after the cathedral city in southeast England. It is biennial, forming a prostrate rosette in its first year and erupting into an inflorescence up to 3 feet (1m) high the following summer. The popular name was originally given to another species, *C. trachelium*, a common perennial in the woods of southern England. *C. trachelium* was once used medicinally, as its scientific name suggests, to cure complaints of the windpipe, or trachea. Thousands traveling along the Pilgrims' Way from the city of Winchester to the tomb of St. Thomas Becket in Canterbury Cathedral—in its time the greatest shrine in Christendom—would have seen *C. trachelium*, with flowers shaped like the bells which hung from their horses' bridles, along the roadside. It is not known when the name Canterbury bells became transferred to *C. medium*. Today's

CUP AND SAUCER
In the "cup-and-saucer" form of Campanula medium (left), the calyx (sepals) have become enlarged and petallike instead of remaining green.

PLAIN PINK
Campanula medium (below) is classic bellflower form, with the bell sitting on a green calyx. The wild species has purplish-blue flowers which occasionally produce pink or white varieties that are easily selected by gardeners.

cultivated Canterbury bells have large blue, pink, or white flowers. The colors are a natural variation of the usual blue, and easy to select out and multiply. Most are the simple bell forms, but a natural mutation, with an enlarged, petaloid calyx, on which the corolla seems to sit, has been selected to form a variation known as "cup-and-saucer."

LOBELIAS: UNTAPPED POTENTIAL

Lobelia is perhaps most familiar in the form of the blue-flowered bedding plant *Lobelia erinus*, which was introduced to northern Europe from South Africa in 1752, and cultivated by Phillip Miller (1691-1771), curator of London's Chelsea Physic Garden. Its use in unsophisticated bedding schemes, often associated with sweet alyssum (*Lobularia maritima*) and scarlet pelargoniums, seems to have turned the gardening snobs against lobelias in general. But it is only one of 365 species (more than there are in the genus *Campanula*, and naturally distributed over a much wider area) representing a surprisingly untapped potential for cultivation. There are tiny creeping lobelias in New Zealand, a submerged aquatic in northern Europe, robust herbs in the Americas and, most bizarrely of all, giant forms in the mountains of East Africa. All are members of the Campanulaceae family, although none of them looks very much like a *Campanula*. A close look at the flower, however, will show the same five-lobed corolla, but which in lobelias forms two lips, the uppermost with two lobes, and the lower with three.

The name *Lobelia* became attached to these plants in honor of Matthias de l'Obel—latinized as *Lobelius*—a Frenchman who became botanist to King James I of England in the 1600s. There seems to be no direct connection between de l'Obel and the plants named after him; the family name is derived from *abèle*—white poplar (*Populus alba*)—and thus passed from plant to man and back to plant.

The first lobelia to become familiar to European gardeners was the perennial *L. cardinalis*—the cardinal flower—a brilliant, scarlet-flowered species introduced from North America in the early 1600s, probably by the French collector Jean Robin "neere the river of Canada, where the French plantation in America is seated," according to his English contemporary, John Parkinson. The gardening writer Jane Loudon (1807-1858) recounted that it had apparently earned its name after Queen

BEDDING FAVORITE
Lobelia erinus (above) is the common bedding lobelia, a trailing annual or shortlived perennial from South Africa.

CARDINAL COLOR
The color of Lobelia cardinalis (left) is similar in color to a cardinal's robes. It is widely distributed in North America and thrives in rich, moist soil, where it can reach 3 feet (1m) or more in height.

Giants of East Africa

In the cold alpine zones of the tropical mountains of East Africa, from Ethiopia to Malawi, grow giant lobelias. From large, slow-growing rosettes of leaves, either sessile or borne on a stem, an inflorescence emerges at irregular intervals. Some flower spikes reach 9 feet (3m) or more in height, while others are squatter and resemble an inverted pineapple. The plants are superbly adapted to their environment; their large bulk permits warmth absorbed during the day to be stored and only slowly lost during the frosty nights. The leaves of the rosettes furl inwards each night, further insulating the growing point from frost. In some, the rosette holds a pool of water, into which a natural antifreeze is secreted, causing the water to freeze at a lower temperature. Such specialized requirements generally make giant lobelias unsuitable for growing elsewhere; but they thrive in San Francisco, where they seem to enjoy the fog.

Henrietta Maria (1609-1669), French consort of the English King Charles I, laughed on seeing it for the first time, and said it reminded her of a cardinal's sock. *L. cardinalis*, like its blue-flowered counterpart, *L. siphilitica*, is a plant of marshland and needs to be cultivated in moist, fertile soil. Pehr Kalm, a protégé of the Swedish naturalist Carl Linnaeus, reported after a visit to North America that the native American Indians used *L. siphilitica* as a cure for syphilis, and although its healing properties were much tried out in Europe—with no recorded success—the name stuck.

The scarlet flowers of *L. cardinalis* and several other New World species of lobelia, including the large *L. tupa* from the coast of Chile, are adapted to pollination by hummingbirds, whose beaks probe into the tubular flowers in search of nectar. Birds, unlike bees, can see the color red, and any flower with a tubular shape and red coloration is more than likely to be bird pollinated, with sunbirds taking over the niche in the Old World where hummingbirds do not occur. The sap of *L. tupa* is narcotic (and possibly hallucinogenic) and was used by Peruvian Indians to treat toothache.

CURE FOR SYPHILIS?
Lobelia siphilitica (above) comes from marshy places in North America, and will cross with L. cardinalis to produce handsome purple hybrids (L. x gerardii).

PLANTS FROM THE DRY SOILS OF
THE MEDITERRANEAN LANDS HAVE
FROM MEDIEVAL TIMES BECOME SOME OF THE MOST UBIQUITOUS AND
ENDURING SPRING AND SUMMER BORDER PLANTS.

Dianthus 'Mrs Sinkins'

PINKS AND WALLFLOWERS
(CARYOPHYLLACEAE AND CRUCIFERAE)

During the reign of Caesar Augustus (43 B.C-A.D. 14), it was reported back to Rome that the inhabitants of Hispania (Spain) flavored their wine by immersing clove-scented carnations, *Dianthus caryophyllus*, in it. The dark, reddish pink carnation grew wild in southern Europe, but as the dunking habit spread, so did the plant, and it was soon cultivated all over Europe, and double variations of the flat wild flower soon appeared. Relics of early cultivation

*North
Atlantic
Ocean*

*North
Sea*

Dianthus plumarius
The wild pink comes from the eastern Alps. Like most Dianthus, it likes a lime-rich soil in full sun and will sometimes become established in an old wall.

Key

Tundra

Boreal coniferous forest

Temperate woodland and grassland

Mediterranean

Desert

Tropical woodland and grassland

Tropical rain forest

Matthiola incana,
ancestor of the fragrant stocks, is widely distributed in the wild along the coastlines of Europe, often growing on cliffs or beaches just above hightide line. It was brought into *cultivation during the 1500s as a bedding plant for early summer display.*

Dianthus knappii

The only wild Dianthus to have yellow flowers comes from western Yugoslavia. Although the wild plant is delicate and modest in appearance, it has become an important source of yellow genes for breeding yellow or orange pinks.

in the north of Europe can still be found today on Norman castles in France and England, where *D. caryophyllus* flourishes in the dry, arid conditions of old walls. In England the flower earned the nickname "sops in wine" or "clove gillyflower," and in German, the word for clove and carnation are the same: *Nelke*. Even the word gillyflower takes us back to the clove theme; it stems from the Arabic *quaranful*, a clove, which became *karyophillon* in Greek, *caryophyllus* in Latin, through *garofolo* in Italian to the French *giroflée*, and English gillyflower. The plant has many common names, which is a measure of its popularity.

"Carnation" itself was first used in 1538, and may be a corruption of the word

THE RANGE
of pinks and crucifers

Crambe cordifolia

An enormous perennial from the Caucasus Mountains which produces branched inflorescences that may be 6 feet (2m) high and covered in thousands of small, honey-scented flowers, and leaves like those of an overgrown cabbage. It is too large for small backyards, but makes a fine specimen plant for large herbaceous borders where a bold effect is desired.

Black Sea

Caspian Sea

n Sea

Malcolmia maritima

The Virginia stock comes from the Mediterranean coast where it enjoys full sun. In cooler climates, it does best at the edges of sunny, well-drained borders.

coronation, or crown, recalling the flower's sacred status and use in ceremonial crowns in ancient Greece. In antiquity the flowers were called *Diosanthos*—"flowers of Zeus" after the father of the gods. Over half of all wild *Dianthus* species today occur in Greece.

The wild carnation was so popular throughout Europe during the Middle Ages, a companion to plants such as *Iris germanica*, lilies (*Lilium candidum*), and peonies, that many named cultivars had been developed by the early 1600s. In England, John Parkinson, apothecary to King James I, grew nearly 50 different clones, with flowers varying in color and size, such as the small 'Master Bradshawe's Dainty Lady', or 'Master Tuggie's Princesse' with white flowers flaked with red, but sadly these are all now extinct. Then they were grown as border plants, but from the mid 1700s it became fashionable to cultivate larger-flowered carnations under glass, to protect the top heavy blooms from inclement weather, and to extend the flowering season.

REPEAT-FLOWERING BLOOMS FROM CHINA

Perpetual flowering was made possible by hybridization with *Dianthus chinensis* in the early 1800s, and this led to the great popularity of large carnations for cut flowers and corsages; carnation growing is now an important industry in many parts of the world. *D. chinensis* was cultivated in China long before its introduction to Europe in 1705, when seeds were sent to the Abbé Bignon (1662-1743), librarian to the French king, Louis XIV, and cultivated in Paris. It is a shortlived species, but its brightly colored, single flowers rapidly became in demand throughout Europe, and double flowers appeared in 1719.

Later introductions of garden varieties from Japan led to the development of improved strains with more compact growth, which are still often used as bedding plants. A feature of all *D. chinensis* cultivars is the darker-colored eye in the center of the flower; a recent, striking development is the cultivar 'Black and White

---

THE CONQUEST OF EUROPE
Pinks and carnations achieved early popularity in the civilized world and gradually became cultivated throughout Europe.

 Pinks

Carnations

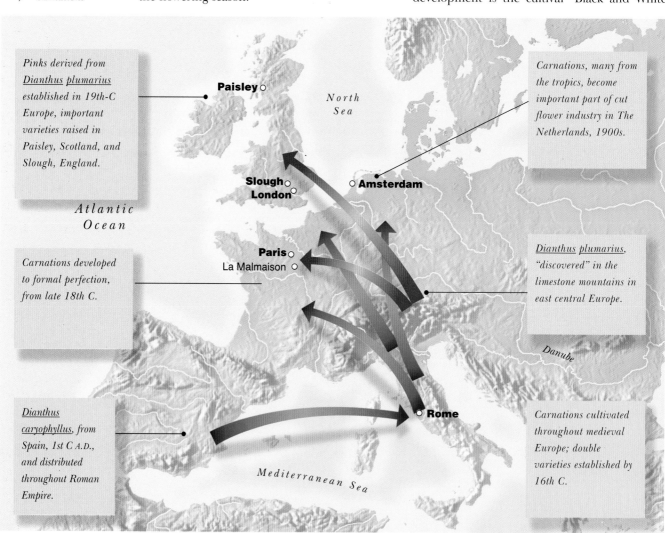

Pinks derived from *Dianthus plumarius* established in 19th-C Europe, important varieties raised in Paisley, Scotland, and Slough, England.

Carnations, many from the tropics, become important part of cut flower industry in The Netherlands, 1900s.

Atlantic Ocean

North Sea

Carnations developed to formal perfection, from late 18th C.

Dianthus plumarius, "discovered" in the limestone mountains in east central Europe.

Danube

Dianthus caryophyllus, from Spain, 1st C A.D., and distributed throughout Roman Empire.

Carnations cultivated throughout medieval Europe; double varieties established by 16th C.

Mediterranean Sea

Paisley　Slough　London　Amsterdam　Paris　La Malmaison　Rome

Minstrels', whose double flowers are deep chocolate purple with a white edge. Unfortunately the quest for size and perfection of shape and color has often led to the loss of scent, which in most florists' carnations is very faint or totally absent.

Dainty relatives

The modern, fragrant border pinks are derived from *D. plumarius*, a native of the mountains of central Europe, which is smaller flowered than *D. carophyllus*. Both the French and English common names for *D. plumarius*—"oeillet" and "pink"—are derived from older words for a small or twinkling eye. It was not until the early 1700s that the flower gave its name in English to the color pink, and later still to the method of

THE CENTRAL EYE

Hybrid descendants (above) of the alpine Dianthus plumarius *retain the characteristic central "eye" which gives the flower its French nickname of* oeillet *(*oeil*—eye).*

CARNATIONS

Centuries of breeding have created the modern carnation (right) but sadly has often lost the scent of the wild species.

cutting a piece of cloth with "pinking shears" to produce a zigzag or "pinked" edge. The specific name *plumarius*, means "feathered," relating to the flower's fringed petals. During the 1700s breeders strove to produce smooth-edged, perfectly round flowers, often with strong contrasts of color. Prominent amongst these breeders were miners and weavers in northern England and Scotland who also produced a cloth known as paisley, after a Scottish weaving town of the same name, in which they tried to reproduce the distinctive pink of the flowers. They would have considered the fringed white petals of 'Mrs. Sinkins', a longtime favorite border pink because of its freedom of flowering, strong scent, and old world charm, ragged and untidy. 'Mrs Sinkins' was introduced in the 1870s by the master of a workhouse in Slough, England, who named it after his wife.

Experimenting with sweet william

The multiheaded sweet william (*D. barbatus*), is a native of southern Europe, from the Pyrenees to the Balkans. It was taken into cultivation by monks as early as the 1100s, and sufficiently common by 1533 to be used for bedding out the gardens of the English king Henry VIII's palace of Hampton Court near London. The origin of the English name is unclear; it seems unlikely to be named for William of Normandy, Conqueror of England—who was anything but sweet, and it is more probable that the name refers to St. William of Aquitaine.

The sweet william was developed by florists in the 18th and 19th centuries, and also used in an early experiment in plant breeding, when Thomas Fairchild (1667-1729), an English nurseryman, tried to hybridize it with a carnation. He had some qualms about the religious morality of such an act, but was successful in producing what became known as 'Fairchild's Mule', a sterile hybrid which was the first deliberately produced hybrid plant.

CRUCIFERS: FLOWERS FOR STONY SOILS

The cabbage family, Cruciferae (now also known as Brassicaceae), contains many weeds and important crop plants such as cabbages, turnips, and oilseed rape, as well as many a fragrant garden plant. What is common to them all is the arrangement of the four petals in the shape of a cross—hence the family name which means "cross bearing."

Wallflowers (*Erysimum cheiri*, formerly *Cheiranthus cheiri*) are symbols of love in

adversity. The metaphor derives from the tale of Elizabeth, daughter of the Earl of March, who was betrothed to the heir to the throne of Scotland and confined to her father's castle. She favored a humbler man however, who, by serenading beneath her window, outlined a plan for elopement. Elizabeth dropped a sprig of wallflower to show that she had understood. Unfortunately, when the time came, she failed to fasten the rope securely and fell to her death. Her mourning lover traveled Europe as a minstrel, wearing (when possible) a sprig of wallflower in his cap.

In their wild form wallflowers are usually yellow, but their cultivated forms were often striped with brownish-red, or completely red. An old double red form known as 'Bloody Warrior' was once greatly prized, but became very scarce earlier this century. Fortunately it has been rescued and is again widely available, a living link with the Elizabethan gardens where it may have originated.

Quick to flower stocks

Stocks are derived from the European seashore plants *Matthiola incana* and *M. sinuata*. Both species occur in single and double forms, and have long been known for their fragrance—they were once called stock gillyflowers. The scientific name of the true stocks, *Matthiola*, commemorates Pierandrea Mattioli (1500-1577), an enterprising Italian physician, who became personal doctor to the Holy Roman Emperor, Maximilian II, but was killed by

A TOUCH OF VELVET
In the wild, wallflowers,
Erysimum cheiri (above),
bear small yellow flowers,
but over the centuries
gardeners have selected for
larger, velvet-textured
variations ranging from
cream to deep blood red,
and bicolored.

SWEET WILLIAM
The scented, multiflowered
heads of Dianthus barbatus
(left) have been cultivated
in gardens large and small
from the1500s. They are
usually biennial, but
annual cultivars have also
been raised.

Show flowers

During the late 1700s a number of plants became the subject of devoted attention by British growers and breeders, who competed to produce flowers that most nearly approached a perfect ideal. They were known as "florists," and they initially concentrated on dianthus, primula auricula, polyanthus, hyacinth, tulip, ranunculus, and anemone. Later pansies and sweet williams were taken up, and the tradition now survives in the breeding of any show flower in which shape and color are important, such as dahlias and chrysanthemums.

The growing of florists' flowers became associated with artisans in industrial cities, who cultivated the plants in their small urban plots. Societies devoted to different flowers were formed, holding annual feasts and show, such as the Paisley Pinks society founded in Paisley, Scotland in 1782, whose members, mostly weavers,

Paisley Pink

grew some 300 cultivars between 1828 and 1850. The flowers were supposed to be perfectly symmetrical, not less than 2 inches (5 cm) in diameter and evenly colored; laced flowers, in which pink or red was superimposed on a white background were the most highly prized, but bolder "black-and-white" flowers were also grown. Some are still popular garden plants. Florists' societies waned by the mid 1800s, when industrialization altered the way of life for formerly self-employed artisans. Diseases such as viral infection may also have been a problem where large numbers of similar plants were grown, causing a general decline in vigor which in turn caused the growers to lose interest.

plague in 1577. His botanical fame rests on his Commentary on the works of the Greek physician Dioscorides, published in 1544.

The wild white and purple stocks (*M. incana*) have been cultivated since 1536. A few years later, a red form was found, and double flowered versions appeared in 1563. From these foundations, gardeners rapidly developed strains of double flowered plants with colors ranging from yellow to deep purple. Double stocks are sterile, so it is necessary to gather seed from single flowered plants and select out the plants that will produce double flowers. Other strains include the "ten week" stocks, so named because they are supposed to flower within ten weeks from sowing, and the Brompton stocks, developed in the London suburb of Brompton during the 1700s, which are bushy biennials with double flowers.

From weed to dried floral display

The first account of *Matthiola sinuata*, a pale, purple-flowered species which has also been used in breeding modern garden stocks, was given by the English gardener and apothecary John Parkinson, who related that it "was brought out of the Isle of Ree by Rochel by Mr. John Tradescant." John Tradescant the Elder (c.1570-1638) was a gardener and nurseryman who worked for several important people, including George Villiers, Duke of Buckingham (1592-1628), favorite of the homosexual King James I of England. Tradescant accompanied Buckingham on what was to be a disastrous attack on the Ile de Ré off the coast of western France, and despite his military duties found time to collect plants.

As its Latin name *Malcolmia maritima* suggests, Virginia stock is a native of seashore locations. It is also a Mediterranean native, and has nothing to do with the North American state of Virginia. The stock has become a favorite with

DOUBLE TAKE
To grow double stocks (right), seedlings must be selected at an early age by inspecting the color of the cotyledons (seed leaves): those which will produce double flowers are pale green, while singles are dark green.

children because of the speed with which it grows from seed to flower. It was introduced to England in 1713, but its familiar name was not acquired until 1770, apparently coined by someone ignorant of the plant's true origin.

Another crucifer from the coastal regions of southern Europe is sweet alyssum (*Lobularia maritima*), an annual or short-lived perennial. Its natural habitat is in dry, sandy places, and it has become a reliable, sweet-smelling, summer bedding plant. Most cultivars retain the white flowers of the wild plant, although selection has

led to the plants becoming more compact. However, in recent years, breeders have established cultivars with red, purple, or even pale apricot flowers.

From weed to dried floral display

Lunaria annua is a native of southern Europe, but has been grown in northern gardens since the mid 1500s. It is enjoyed both for its spring flowers and the later fruits which are often dried for winter decoration. The plant's specific name of *annua* is misleading, for it is a biennial, flowering in the second year after germinating. Purple or white flowers are followed by round, flattened fruits about 2 inches (5cm) long. They have two outer walls (technically valves) that surround and initially protect the septum on which the seeds develop, but as the seeds ripen the valves drop off, leaving the silvery, translucent septa behind. It is these remarkable structures that inspire the plant's names: *Lunaria*, suggesting the moon, the American "silver dollar," and the French *herbe aux lunettes*—"spectacle flower."

*T*HE LARGEST FAMILY IN THE
PLANT KINGDOM EMBRACES AN
ENORMOUS DIVERSITY OF RAYED FLOWERS FROM DAISIES AND EDELWEISS
TO ASTERS, DAHLIAS, AND CHRYSANTHEMUMS.

The common daisy

*T*HE *D*AISY *F*AMILY

(COMPOSITAE OR ASTERACEAE)

*T*he Daisy or Aster family is represented by more than 21,000 species throughout the world, ranging from tiny herbs to woody-stemmed giants. All share a distinctive "composite" flowerhead made up of individual flowers of two different types. Members of the

Leucanthemum maximum, *a large, white-flowered perennial daisy, is indigenous to the alpine meadows of the Pyrenees in France and Spain. It grows* *well in fertile soil but is less often seen in gardens than its descendant L. x superbum, the Shasta daisy, which is even larger and more vigorous. Regular division will keep them in bounds. As with many composites, they make attractive cut flowers, lasting well in water.*

A COMPOSITE WORLD
—*the family range*

Solidago canadensis, *or golden-rod, is a large, summer-flowering border plant from woodland edges and rough fields in eastern North America. Its bright yellow flowers are a much visited source of pollen and nectar for bees and other insects. Introduced to Europe in the 1600s it is now widely naturalized there.*

Gazania hybrids, *with their magnificent, bright flowers in shades of white to coppery-red, are derived from G. rigens and G. pectinata, which make a major contribution to South* *Africa's spring displays of wild flowers. Although perennial, they can be successfully grown as summer annuals in cold climates, and selected individuals can be perpetuated by means of cuttings. The genus takes its name from the medieval scholar Theodore of Gaza (1398-1478).*

family are often known simply as composites. In 1871 the gardening writer William Sutherland referred to daisies and other composites as a "horde of barbarians which no sane gardener would admit within the boundaries."

Typical daisy flower

The common daisy *Bellis perennis* certainly must be one of the most familiar—but often least desired—sights in the gardens of temperate lands. This small plant with prostrate (flattened) rosettes of leaves originated in northern Europe but has spread to many regions, including northeast North America.

The Latin name is derived from *bellum*, meaning "war," as the flowers were said to staunch bleeding from the wounds. The expressions "lie beneath the daisies" and "push up the daisies" refer to the presence of the flowers in the turf over graves. In English folklore, spring is said to have arrived when it is possible to tread with one step on seven—or

more, depending on region—daisies at a time. The common name "daisy," applied widely to plants throughout the family, is said to be a corruption of "day's eye," referring to the flower's habit of opening in the morning and closing at night. Daisies are also used to help a doubting lover ascertain the feelings of a partner, as she (or he) pulls off the outer white florets one by one, reciting: "He loves me, he loves me not …"

Variations on a theme

The flowers of *Bellis perennis* are typical of the composite flower structure, an understanding of which provides a key to how it has been selected and bred. Technically, what seems to be a single

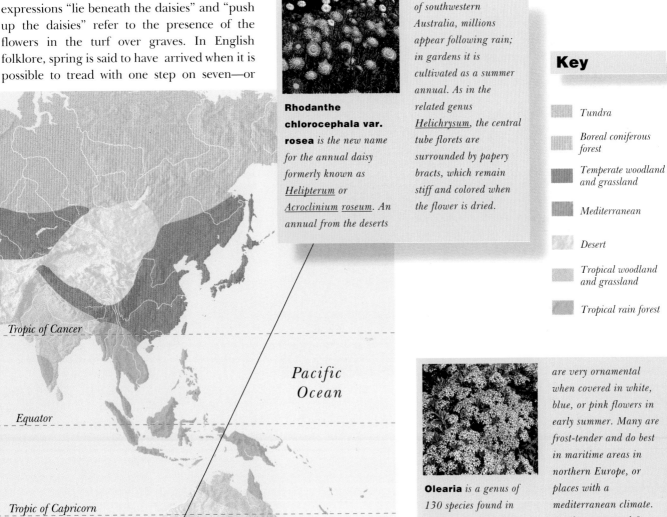

Rhodanthe chlorocephala var. rosea *is the new name for the annual daisy formerly known as* <u>Helipterum</u> *or* <u>Acroclinium</u> <u>roseum</u>. *An annual from the deserts of southwestern Australia, millions appear following rain; in gardens it is cultivated as a summer annual. As in the related genus* <u>Helichrysum</u>, *the central tube florets are surrounded by papery bracts, which remain stiff and colored when the flower is dried.*

Tropic of Cancer

Equator

Pacific Ocean

Tropic of Capricorn

Olearia *is a genus of 130 species found in Australasia from New Guinea to New Zealand. Although a few can become quite large trees, most are small, evergreen shrubs which are very ornamental when covered in white, blue, or pink flowers in early summer. Many are frost-tender and do best in maritime areas in northern Europe, or places with a mediterranean climate. Easily propagated from cuttings, they usually grow rapidly when planted out. The picture shows* <u>Olearia</u> x <u>haasti</u>, *an artificial cross.*

Key

Tundra

Boreal coniferous forest

Temperate woodland and grassland

Mediterranean

Desert

Tropical woodland and grassland

Tropical rain forest

completely absent, and in others, all the florets are rayed, giving a double-petaled appearance. Such doubling is a feature of many Compositae garden cultivars, and in some cases the double version has almost supplanted the single wild form. Double forms of *Bellis perennis* have been cultivated since the 1500s. They are selected from variants in which the inner tube florets have become rayed, like the outer ones, and changed in color to red or pink.

The remarkable variant *B. perennis* 'Prolifera', also known as the "hen-and-chickens daisy," has smaller subsidiary flowers around the main flower at the top of the stem. Presumably it arose as a wild mutation, and still requires an open habit, fertile soil, and regular divisions to stay healthy.

OX-EYE DAISIES

Members of the composite genus *Leucanthemum*, meaning "white flowers," are also called daisies. These summer-flowering perennials, yellow-centered with a white surround, were all originally natives of Europe but are now common in many areas, including North America. One example is the ox-eye daisy *L. vulgare*, a robust garden and meadow plant that thrives in exposed places. In Britain the ox-eye daisy is also known as the moon daisy, from its white shimmer at night.

Other leucanthemums have also traveled abroad, such as *L.* x *superbum*. This is widely known as the "Shasta daisy," after Mount Shasta and its nearby city in northern California, U.S. However the origin of this name is obscure, because it actually originated in northern Europe. Double versions such as the all-white 'Esther Read' are in cultivation.

NEW WORLD ASTERS

Michaelmas daisies are important fall-flowering composites. They bloom at the time of the Christian feast of St. Michael, or Michaelmas Day—September 29 in the Western calendar. However this name was only applied after the Western World adopted the Gregorian Calendar in 1752. St. Michael's Day then fell eleven days earlier, and so coincided with the peak flowering period of the plants.

Michaelmas daisies are derivatives of the wild species *Aster novi-belgii*, *A. novae-angliae* and related species from eastern North America. *Novae-angliae* is clearly "New England." But *novi-belgii*, "New Belgium"? In fact this refers to the Dutch settlement of New Netherlands,

flower is, in fact a whole head of florets which is called the capitulum. The outer ring consists of ray florets which are white, sometimes tipped with red. Each of these is a minute yet complete flower made up of a single petal and reproductive organ.

The daisy's yellow center is made up of a disk or bunch of tube florets. Again, each one of these is a single flower, with tiny but perfect reproductive organs, but in this case, no petals.

The color contrast between the outer ray florets and inner tube florets—where both are present—is found throughout the Compositae family. But in some daisies, ray florets are

SMARTENED UP
The cultivated version of the common daisy, Bellis perennis, has a double form because each of the central tube florets of the composite flower has developed a ray, which may be quill-like, as above, or petal-like.

founded in 1623 at the mouth of the Hudson River. It was overrun by the British in 1684, when its capital, New Amsterdam, was renamed New York.

Rascally plants

Both original wild species of Michaelmas daisy are erect, herbaceous plants, bearing numerous flowers with outer ray florets in shades of pink, purple, and blue, and yellow central tube florets. They were introduced to Europe in the early 1700s. Hybrids form easily in the genus *Aster* (the Greek word for a "star"), both in the wild and in cultivation—thereby confusing

VIBRANT COLORS
The late summer flowers of Aster novae-angliae *'Andenken an Alma Potschke' are intense and vivid of color; they need to be thoughtfully sited to avoid clashes.*

taxonomists. Great American botanist Asa Gray (1810-1888) was moved to declare: "Never was there so rascally a genus; they reduce me to despair." Yet this propensity has led to the continuing development of innumerable fine hybrid Michaelmas daisies, which are well suited to temperate gardens.

Most of these cultivars are larger and brighter in bloom than their wild relatives. For example, the wild *A. novae-angliae* has small violet-purple flowers, but many derived cultivars are much larger, and vary from white to deep red. Among the most striking is *A. novae-angliae* 'Andenken an Alma Pötschke', a startling

border plant with intense pink flowers, which was released by a German nursery in 1969.

DUAL-PURPOSE SUNFLOWERS

In North America, the only important crop plant developed by early agriculturalists was a composite—the sunflower, *Helianthus annuus*. This complex hybrid is not known in the wild state and was presumably bred in ancient times. Remains of its seeds have been found at Amerindian sites up to 3,000 years old. It spread southwards to Central America and Peru, where it became revered as an emblem of the Sun God and was used as a motif in temples. The plant remained an important crop for native Americans until European settlement, from the 1500s onward.

The sunflower's generic name is straightforward, from *helios*, meaning "sun,"and *anthos*, "flower." This is suggestive not only the flowerhead's sun-like appearance, but also its habit of rotating in the direction of the sun through the day: the French and Italian names —*tourne-soleil* and *girasole*—both literally translate as "turn sun."

Such a conspicuous flower—even the early types had enormous heads—immediately drew the attention of Europeans exploring the Americas. The first seeds were sent to Madrid Botanic Garden in 1510. The plant was spread rapidly through Europe, impressing gardeners by its speedy growth, great height, and huge flower heads. By the early 1600s, tales circulated in Europe about sunflowers of gigantic proportions. And the sunflower, plentiful producer of vegetable oil, is the State flower of Kansas, noted for its production of mineral oil.

Boosts to popularity

Tsar Peter the Great (1672-1725) introduced sunflowers to Russia as part of his campaign to modernize his domain. Their seeds were sold on the streets as popular snacks. Later, sunflower oil was used for cooking, and Russia and other CIS countries (formerly the Soviet Union, USSR) are still the largest producers of sunflower oil.

Sunflowers are among the few commercially important plants to be valued equally in the ornamental garden and farmer's field. Many cultivars have been developed, including those with double flowers, depicted in Vincent Van Gogh's painting *Sunflowers*. Cultivars with red- or brown-petaled flowers are derived from a chance find in 1910 of a wild plant growing near

the town of Boulder, Colorado. The sale of Van Gogh's painting in the 1980s, commanding a then world record sum for a work of art, led to a resurgence of interest in the sunflower as a backyard plant and the creation of many new cultivars. Some cultivars produce no pollen, and are therefore especially suitable low-allergy candidates for the cut-flower market.

Other species of *Helianthus* are also valued in gardens for their late summer and fall displays. Most, such as *H. decapetalus*, are perennials, with yellower flowers than *H. annuus*. They include the double-flowered *H. decapetalus* 'Loddon Gold' and the semidouble 'Flore Pleno'.

One member of *Helianthus* is better known by another name—*H. tuberosus*, the Jerusalem artichoke. It is grown as it was in pre-Columbian North America, for its edible tubers, although it rarely flowers in more temperate climates. The curious English name is a corruption of the Italian word for sunflower, *girasole*.

DISCOVERING DAHLIAS

Dahlias were first cultivated by the Aztecs in Central America prior to the Spanish conquest in the 1500s. They were first brought to the notice of European naturalists by Francisco Hernandez. His book *History of Mexico* was prepared in the 1570s, but suppressed by the Spanish Government until 1651.

Hernandez records that the plants were cultivated for their tubers and eaten as food by the Aztecs—in the more fertile areas of Mexico. The Aztecs called it *cocoxochitl*, "water pipe". The tree dahlia *Dahlia imperialis* has long, hollow stems, which were used as pipes to carry water from springs and streams to their dwellings.

Swift rise to fame

Dahlias were not introduced to Europe until 1789, when seeds were sent from the Botanic Gardens of Mexico City to the Abbé Antonio Cavanilles (1745-1804), keeper of the Royal Gardens of Madrid. Cavanilles bestowed the scientific name *Dahlia* in honour of Anders Dahl (1751-1789), a pupil of the Swedish naturalist Carolus Linnaeus. These first species, *Dahlia pinnata* with semidouble, purple flowers, and *D. coccinea* with single, red flowers, did not thrive in European conditions. Cavanilles hoped that their swollen roots would be a useful food source. However the verdict was that the tubers were "edible, but not agreeable," being spurned even by cattle. In 1804 the German polymath

SENSATIONAL FIND
The large flowers of the sunflower, Helianthus annuus (left), were a sensation when first introduced to Europe in the 1500s. They show the classic composite structure of central tube florets surrounded by ray florets.

ALTHOUGH NO MEDIEVAL GARDEN HAS SURVIVED, IT IS POSSIBLE TO PIECE TOGETHER A PICTURE OF A TYPICAL GARDEN IN THE DARK AND MIDDLE AGES.

THE MEDIEVAL GARDEN

Modern medieval-style herb garden.

The fall of the Roman Empire in the 5th century A.D. swept the horticultural expertise of Europe into a few corners during the Dark Ages, from which it cautiously emerged many centuries later.

It was the stability of the Carolingian Empire under Charlemagne (742-814) and his successors that enabled horticulture to revive throughout western Europe. Surviving Classical ideas were incorporated into contemporary practice, while the diplomatic relations fostered by Charlemagne meant that new plants arrived from the Islamic gardens of Spain and Turkey.

Records show that royal estates and monasteries were the prime source of gardening. A decree, the *Capitulare de villis*, issued by Charlemagne, listed 89 species of culinary or medicinal plants to be grown on royal estates throughout the empire. The list, which included named cultivars of fruit trees,

may have been compiled by Abbot Benedict of Aniane in southern France, who traveled widely, even reaching India. Another gardening monk was Walafrid Strabo, Abbot of Reichenau, whose poem *Hortulus* (Little Garden) is an aesthetic appreciation.

Paintings of ornamental gardens in the Middle Ages often show a small, secluded plot with trees and flowers in which people are taking their leisure. This exemplified the *hortus conclusus* (enclosed garden), whose inspiration comes from the biblical *Song of Solomon* (4:12-14), "A garden enclosed is my sister, my spouse... Thy plants are an orchard of pomegranates, with pleasant fruits...with all the chief spices." Often interpreted as an allegory of the Virgin Mary, such gardens were filled with flowering plants. These were mostly wild flowers from the fields and woods nearby, but also included rarer exotics.

THE PERFECT PLAN
Extract from the St. Gall plan (right) prepared for Gozbert, abbot of St. Gall monastery, Switzerland, 816-836. It incorporates all the features of the ideal monastery garden: strictly practical with narrow beds divided by paths, for the cultivation of vegetables or medicinal herbs in blocks of a single variety. Even the monks' cemetery doubled as an ornamental orchard.

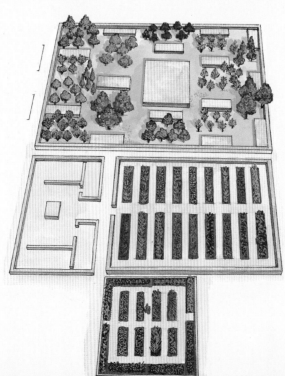

"The Mary Garden" (right) by an unknown Rhenish master, from 1420, shows an idealized and deeply symbolic Medieval enclosed garden. Among the "flowery mead" with wild flowers growing through the grass are white Madonna lilies representing the purity of the Virgin Mary, red roses for Divine love, regal irises for Jesus' royal descent from the House of King David, and cherry for the joys of Heaven.

Alexander von Humboldt (1769-1859) sent seeds of another dahlia species, *D. variabilis*, from Mexico to gardens in Paris and Berlin. They grew to reveal semidouble flowers in pale purple, pink, and yellow, and bred successfully. Only four years later, 55 cultivars were growing in the Berlin Botanic Gardens. By 1815 all the familiar shapes and colors of modern dahlias, except the cactus variant, were established.

The dahlias and the Empress

During the early 1800s, dahlias rapidly became popular across Europe. In France, Napoleon's Empress Josephine is said to have planted them herself at her retreat, La Malmaison near Paris, around 1810-13. Her gardens became a center for their cultivation, where they rivaled her beloved roses for attention.

A possibly apocryphal tale recounts how Josephine jealously kept her dahlias confined to Malmaison until a lady-in-waiting, by the devious means of bribing a gardener, acquired some. The lady unadvisedly boasted of the acquisition. When Josephine heard, she banished the lady, sacked the gardener, and destroyed the entire dahlia collection.

Dahlias reached England at the end of the Napoleonic Wars (and Josephine's life) in 1814. They rapidly became the most fashionable plants in the country, developing into a florist's flower that rivaled the older carnations and pinks (see p. 51).

The spiky flowered cactus dahlias have become distinguished show flowers. They are said to resemble various night-blooming cereus cacti of central America and the Caribbean. The spikiness is caused by the petals curling over along their edges. Cactus dahlias are derived from *D. juarezii*, introduced into Holland in 1872, and named after the then-President of

Mexico, Signor Juarez. The original plant was the sole survivor of a consignment of tubers sent from Mexico; the rest perished on the voyage.

One of the most striking dahlias is 'Bishop of Llandaff', with brilliantly scarlet single flowers above dark purple foliage. It was raised among a batch of mixed seedlings in Cardiff, Wales, during the 1920s by nurseryman Fred Treseder.

CHRYSANTHEMUM: IMPERIAL FLOWER

The chrysanthemum, whose name literally means "golden flower," has been cultivated and treasured in China for nearly 2,500 years. It was praised by the Chinese philosopher Confucius (551-479 B.C.), who particularly admired a dark-flowered variety with dark leaves.

Chrysanthemums did not arrive in Japan until the 4th century A.D., but they soon underwent a further flurry of developments, including flowers of different shapes and textures. Whereas the Chinese preferred a flower with incurved petals, the Japanese favored flattened, less formal blooms. Specialist chrysanthemum shows were being held in Japan by the 10th century A.D. About this time a prized single flower with 16 symmetrical, petal-like ray florets was chosen as the personal symbol of the Emperor, or Mikado. Centuries later, in 1871, it was adopted for the Imperial Standard.

Europe loses out

Most varieties of chrysanthemums were developed in China and Japan, long before Europeans set eyes on the plant. In Japan, especially, their cultivation was limited to gardens of the Emperor and nobility. It is, perhaps, not surprising that the first specimens to reach Europe were of poorer quality. In about 1689 they were grown for a short time in Holland—the Dutch being the only European nation permitted to trade with the Japanese. But these lines soon died out.

The 1700s saw more attempted introductions. The only one to succeed was a purple-flowered cultivar imported by Captain Blancard of Marseilles in 1789. One of these plants reached Kew Gardens near London, England by 1790. In 1796, it was exhibited in large numbers by British nurserymen, and became known as *Chrysanthemum morifolium*—a main ancestor of many of today's garden and pot varieties.

Over the subsequent years, more chrysanthemums arrived in Europe. John Reeves, the British East India Company's tea

IMPERIAL FLOWER
Chrysanthemum cultivars (left) are now an important cut flower, but were originally weedy plants of disturbed ground in central China.

TASTEFUL DAHLIA
Dahlia 'Bishop of Llandaff' (right), which arose in Wales in the 1920s, combines rich red flowers and dark bronze leaves. The frost-sensitive tubers need protection in winter.

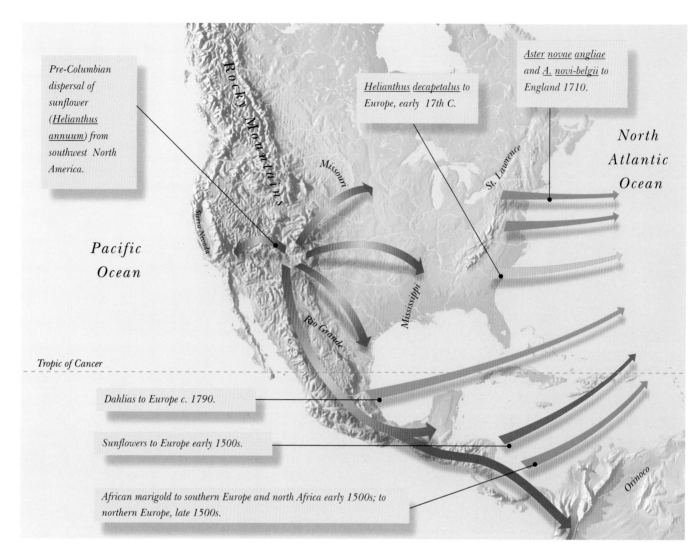

Pre-Columbian dispersal of sunflower (*Helianthus annuum*) from southwest North America.

Helianthus decapetalus to Europe, early 17th C.

Aster novae angliae and *A. novi-belgii* to England 1710.

Pacific Ocean

Rocky Mountains

Missouri

Sierra Nevada

St. Lawrence

North Atlantic Ocean

Río Grande

Mississippi

Tropic of Cancer

Dahlias to Europe c. 1790.

Sunflowers to Europe early 1500s.

African marigold to southern Europe and north Africa early 1500s; to northern Europe, late 1500s.

Orinoco

inspector in Canton, China, was instrumental in acquiring some of the best cultivars from his local region. Between 1821 and 1826 Joseph Sabine (1770-1837), Secretary of the Royal Horticultural Society, described 68 distinct cultivars. He reported to the society that one was known to the Chinese as "Drunken Lady"—"on account of its rosy hue."

Gradually European breeders began to develop their own chrysanthemums. In 1846, Robert Fortune introduced two miniature varieties from the island of Chusan, off southern China, which helped to create the pom-pon chrysanthemums. Another important development was the selection of early-flowering cultivars by the French nurseryman Monsieur Delaux, during the 1880s and 90s. He aimed to produce cultivars that would flower outdoors without damage by frost. Like many composites, chrysanthemums display a strong response to the ratio of daylight and darkness within each 24 hour period. They produce flowers only when daylight begins to shorten,

NEW WORLD DISCOVERIES
The map illustrates how many familiar composites were discovered by early European travelers in North and Central America and taken by them to Europe.

after the summer solstice. Many chrysanthemum require several nights of at least 14 hours' darkness to initiate bud development. The early-flowering types do the same, but mature their buds more rapidly than others after this initial dark-period stimulus has been received and so escape frost damage. This darkness requirement is manipulated by commercial growers, who use artificial light-dark regimes in covered glasshouses to ensure year-round supplies of cut flowers.

Name confusion
Due to a convoluted history, today's garden and florist chrysanthemums are well known to many botanists and gardeners, not as *Chrysanthemum*, but as complex hybrids derived from several Asian species of the related *Dendranthema* x *grandiflorum*. They are all perennial, and suited to pot or garden, depending on treatment. However a more recent decision renames the genus as *Chrysanthemum*—a name that already exists, not for garden chrysanths, but for a few

annual cornfield weeds. This means the weedy former members of *Chrysanthemum* are looking for a new name.

EDELWEISS: ALPINE CHALLENGE

The edelweiss, *Leontopodium alpinum*, is indelibly associated in the West with the European Alps. Its "flower" is even more complex than a typical composite. It is made up of several small, rayless capitula surrounded by woolly white bracts (whose texture give rise to the common English name of "flannel flower."). The genus name means "lion's foot."

The flower is common in alpine meadows and also relatively easy to grow in temperate areas. Like most alpines it requires full sunlight, good drainage, and a cold winter rest period. During the 1800s, a legend arose that the edelweiss was a cliff dweller that needed special courage and skill to acquire from the wild. This encouraged intrepid male lovers to climb and search for specimens, to present as a sign of devotion to their girlfriends. Casualties occurred as the ill-informed and ill-equipped fell off cliffs in the European Alps. In response, and to help its survival, the Tyrolean and German Alpine Clubs imposed fines on their members for picking the flowers. This farsighted move seems to have been the first ever legislation for plant protection.

Other members of the genus are found in the cold, high deserts of Central Asia or in the Himalaya Mountains. In Nepal grows the striking golden edelweiss, *L. monocephalum*, with yellow-tinted hairs and low, mat-forming habit. It is common among the high rocks and meadows, but it has not been successfully introduced to cultivation—yet.

Lasting color

The "everlasting flower," *Helichrysum*, are closely related to the edelweiss. Members of this large genus are found mostly in Africa and around the Mediterranean, although some species occur in Australia. Like the edelweiss in the European Alps, *H. meyeri-johannis* in Africa decorates the hats of successful climbers on Mt. Kilimanjaro. Its large white and red flowers, as the common name suggests, retain their shape and color almost indefinitely. It is named after Hans Meyer, a German Professor of Geography who, in 1889, became the first person to reach the summit of Kilimanjaro.

The familiar yellow, orange, and red everlasting flowers used in dried flower arrangements are derived from the Australian *H. bracteatum*. This wild perennial, shrubby and with yellow flowers, has been bred as an annual for northern gardens, with a variety of yellow to brown flowers. The dwarf cultivar 'Bright Bikini' has red, pink, and orange blooms. Another member of the genus is *H. italica* from the southern European Alps. This has insignificant yellow flowers and is often called the "curry plant" as its leaves have a strong, spicy aroma, but they are not edible.

ORIGINAL CONFUSION
The name "African marigold" was applied to large-flowered cultivars of Tagetes erecta *(left) in the late 1500s, when seeds arrived in northern Europe from North Africa; the plant is actually native to Mexico.*

*T*OUGH LEAVED, WOODY-STEMMED

SHRUBS AND TREES FROM COOL,

WILD HEATHLANDS AND MOUNTAINS AROUND THE WORLD, RANGING FROM

TINY ALPINE HEATHS TO BUSHY AZALEAS AND STATELY RHODODENDRONS.

Rhododendron 'Konigin Emma'

*T*HE *H*EATHER *F*AMILY

(ERICACEAE)

Key

▨ *Tundra*

▨ *Boreal coniferous forest*

▨ *Temperate woodland and grassland*

▨ *Mediterranean*

▨ *Desert*

▨ *Tropical woodland and grassland*

▨ *Tropical rain forest*

Almost all members of the Ericaceae family need acidic or lime-free soil to thrive. Their very presence is a reliable indicator of the type of soil and the rocks beneath, both in the wild and in cultivation.

Heathers and rhododendrons are found usually in cool mountain or temperate regions throughout the world, from the lower slopes of the Himalayan mountains to the open moorlands of northern Europe. In Anglo-Saxon times these barren lands of northern Europe were known as *hoeth*, from which the modern names for both the landscape—heath—and the dominant plants growing there are derived. (The word Erica is derived from the Greek *ereiko*, to break, since drinking an infusion of the leaves was reputed to break up bladder stones).

The dominant plants of heathland are the members of the genera *Erica*—the true "heaths"—and the monospecific genus *Calluna*, or "heather." Both are prime examples of the woody-stemmed, shrubby growth characteristic of the family as a whole. Also characteristic are the distinctive flowers with petals fused into a bell- or funnel-shaped, five-lobed corolla. *Calluna vulgaris* differs from the true heaths, however, because it has an enlarged calyx—a funnel or bell of what appear to be fused petals, but which are really sepals—concealing the small corolla of true petals within.

Kalmia latifolia, *the mountain laurel or calico bush, is the State flower of Pennsylvania. It is a small tree or large shrub with evergreen leaves resembling those of a Rhododendron, and produces bell-shaped flowers in late spring.* Kalmia *flowers are usually mid pink, but can vary from crimson to almost white, and all have stamens arranged like umbrella-spokes against the corolla. Other, smaller, species occur elsewhere in North America, thriving in cool mountain areas on acidic soil.*

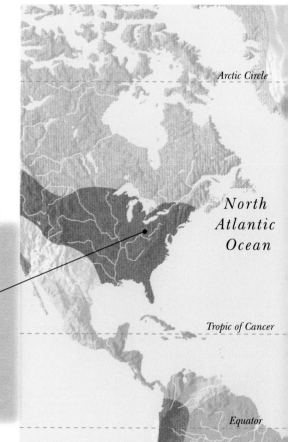

Arctic Circle

North Atlantic Ocean

Tropic of Cancer

Equator

NOT SO COMMON HEATHER

Calluna vulgaris is often called the "common" heather. It covers large areas of northern Britain, and also scattered sandy regions in the south, coloring the hills and mountains purple during its summer flowering season. But it is far from common elsewhere in Europe. It did, however, make the journey to the New World. Emigrant Scots took seeds to Canada and New England to remind them of their highland home, and heather has since naturalized there. Cultivars have found their way into acid cultivated soils around the world.

There is only one species of *Calluna*, but it is extraordinarily variable in the wild and in cultivation, ranging from tight, compact mounds, to erect, open shrubs, and flowers ranging from typical pinkish-purple to cultivars in bright red or white.

Heather was originally an understory species of open forest, but since the large-scale felling of woodlands, it has adapted to the harsher conditions of open heath and moorland. It is actively encouraged by landscape management in many areas since the young shoots and buds, especially, are the favored food of an important game bird, the red grouse *Lagopus lagopus scoticus*.

Practical use and lucky charm

Until the 1700s in the Scottish Highlands, tips of heather shoots were mixed with malt and brewed to produce heather ale. The practice waned but has recently been revived, and the ale is commercially available once again. White

Arbutus unedo, *the strawberry tree, has bright red, strawberry-like fruits which are dull and insipid to taste. They take a full year to reach ripeness, flushing red just as the white flowers open in fall. This small evergreen tree survived the ice ages in the mild conditions of western Europe and the Mediterranean basin, while related species are found in North America.*

Rhododendron arboreum *is the national flower of the Himalayan kingdom of Nepal, where it flowers in May—the sight of snow-capped peaks framed by its bright red, pink or white flowers is unforgettable. A large tree which may reach 100 feet (33m) in the wild, it seldom exceeds 40 feet (13m) in cultivation. It is also found in western China, and is an important parent of hybrid rhododendrons.*

THE FAMILY RANGE AROUND THE WORLD

Erica carnea, *the winter-flowering heath, is found in the European Alps where it blooms in late winter and early spring. In cultivation in milder climates it is often in bloom in midwinter. Unusually for this family,* E. carnea *does tolerate some lime in the soil, but too much causes the foliage to become yellow. The cultivar illustrated is 'Springwood White' which was spotted in the 1920s and grows into a large, weed-smothering dense mat.*

TREE SIZED HEATHER
Erica <u>*arborea*</u> *(above)*
reaches tree proportions
under the ideal conditions
of an almost frost-free
climate. It is usually seen
as large bush that can
make a striking
counterpoint to the
typically prostrate ericas.

heather is selected and cherished as a symbol of good luck in many countries where Scots have settled. This is possibly a reflection of the original scarcity of white-flowered individuals among the millions on the Scottish moors. Or the reason may be Clan MacDonald's success in battle, in 1544, attributed to wearing sprigs of white heather in their bonnets.

The belief in lucky heather was ardently fostered by the English Queen Victoria who insisted on its inclusion in the wedding bouquets of her daughters. Today a variety of travelers, and gypsies peddle sprigs of "heather" in many cities, although the plant is often in fact, a completely different dried shrub such as *Limonium*, sea lavender (from the family Plumbaginaceae).

ERICAS: THE TRUE HEATHS

The pink- and purple-flowered species of *Erica* are characteristic plants of European heathland. But more than nine-tenths of the 735 species in the genus are actually found in South Africa, with more than 500 endemic to the Cape area. Such an extraordinary diversity of closely-related species in a limited area is unparalleled anywhere in the world.

The Cape heaths are typical of the *fynbos* shrubland, a relatively dry, well-drained landscape dotted with grasses, low to medium height shrubs, and the occasional tree. From the original small, pale flowers of the genus, many of the wild Cape ericas have evolved long tubular flowers in shades of white, yellow, and bright red. Some of these were imported for greenhouse cultivation in Europe during the 1800s. But they mostly died out during the First World War (1914-18), probably through lack of labor to maintain the greenhouses and the specialized requirements of these choosy plants. Today, very few South African heaths are cultivated outside their native region.

Height variation

Other *Erica* species are found in many northern gardens, often in a rock garden where the soil is maintained in acid condition, but they do not like shade.

Most heaths are low- to medium-sized shrubs, but in southern Europe and throughout Africa, and Madagascar, some species reach tree dimensions. The slow-growing tree heather, *Erica arborea*, is found from the European Alps south to the East African mountains. It grows to 25 feet (8m) high and often develops woody knobs and burrs at its stem bases. These are the traditional source of the briar wood used to make tobacco pipes. Briar is a corruption of the French word for erica, *bruyère*. The hardier variety *E. arborea alpina*, from the mountains of Spain, is valued for its height in a predominantly low-level heather garden, and bears profuse white flowers in spring.

Erica carnea, the winter-flowering heath, is a low, mat-forming species of great hardiness from the mountains of central Europe. It is valued for its white, pink, or red flowers produced in midwinter, and is almost unique in the genus for its tolerance of lime in the soil. The vigorous cultivar 'Springwood White' has prolific white flowers and derives from plants

HEATHER LANDS
Calluna <u>vulgaris</u>, *the common heather (right), covers large areas of upland Britain with its purple flowers each summer; gardeners have selected numerous color forms that range from white to crimson.*

The man who made rhododendrons popular

Few botanists have been so influential as Joseph Dalton Hooker (1817-1911). His father, Sir William Jackson Hooker (1785-1865), was director of Kew Gardens near London, England, so the young Hooker's path in life was well signposted. He started his botanical career in 1839 on a four-year expedition to the Antarctic Ocean, gaining firsthand experience of the flora in many regions of the southern hemisphere. A few years previously, a young naturalist named Charles Darwin had begun to develop his theory of evolution by natural selection. Joseph became one of Darwin's principle confidants, and was copresenter of Darwin and Wallace's scientific papers on evolution in 1858.

On his visit to India, from 1847 to 1851, Joseph laid the foundations for his major published work, the seven-volume *Flora of British India* (1872-97). During his trip to the borderlands of Sikkim, he and his companion, Archibald Campbell, were  imprisoned by hostile Tibetans. They endured six weeks of solitary confinement before the authorities negotiated their release. Despite such problems, Joseph collected 30 new species of *Rhododendron*, both as herbarium specimens and seeds. They were described in his magnificent book, *Rhododendrons of the Sikkim-Himalaya* (1849-1851). The book, specimens, seeds, and resulting seedlings, distributed from Kew, established the great popularity of rhododendrons in Britain.

Hooker was President of the Royal Society from 1873 to 1878, received a knighthood in 1877, and continued work on his beloved *Impatiens* to within a few days of his death.

69). This opened the floodgates for rhododendron enthusiasts. During the next century, a torrent of new species and varieties emerged from China and the Himalayas, bound for the gardens of Europe and North America.

Missionary zeal

During the 1800s, China was a remote nation, far more so than today, and closed to most Westerners. Among the few foreigners to penetrate its interior were French priests, employed by the *Missions Etrangères* as medical missionaries. They aimed to set up hospitals in remote places, but some of these pioneers had more than a passing interest in biology, and especially botany.

Perhaps the most important and active in this respect was Père David (1826-1900). He brought the giant panda to the attention of Western science, and discovered another species that now bears his name, Père David's deer. He was also an active botanical explorer from 1860 to 1874. In 1869 he spent several months in the principality of Mupin, on the Sichuan-Tibet border, exploring the mountains with great thoroughness. This was the first extended and systematic study of the flora of a Chinese district by a European botanist. It also led to the naming of a new genus in his honor, the dove, or pocket-handkerchief tree, *Davidia involucrata*.

During his survey of the Mupin area, Père David discovered 16 species of rhododendron which were new to science. They included the now-familiar bristly species *R. strigillosum* with deep red flowers. But he collected very few specimens or seeds. Most of his discoveries were introduced to cultivation later, by Anglo-American planthunter Ernest Wilson.

French sacrifices

Another 19th-century French priest, Père Jean Marie Delavay (1834-1895), was stationed in China for 28 years. Despite his missionary duties, he managed to collect, prepare, and label an astonishing 200,000 herbarium specimens, representing 4,000 species—of which 1,500 were new to science. Delavay worked mainly in northwest Yunnan, a region of outstanding botanical richness which continues to attract planthunters today. Many of the rhododendrons he described are now established horticultural favourites, such as *R. fastigiatum*, a low shrub with blue or purple flowers, and the tree-sized *R. irroratum* with pink or yellow flowers. Unfortunately, few of his

AMERICAN INFLUENCE
The deciduous azaleas (above) with their characteristic, open, upward-facing flowers, are a complex group derived from many European, Asian, and North American species of rhododendron.

CHINESE ORIGINS
Rhododendron irroratum (right), which comes from western China and southeast Asia, has the bell-shaped flowers typical of most of the large-flowered Asian rhododendrons.

found during the 1920s on Monte Carreggio in the Italian Alps by a Mrs Walker from Stirling, Scotland. She collected cuttings and cultivated them at her home, Springwood, after which the cultivar was named.

Erica erigena, a close relative of *E. carnea*, is a shrubby, erect species from southwest Europe. It reaches 6 feet (2m) in height and tolerates a wide range of conditions and soils, even slight saltiness. The hybrid *E.* x *darleyensis*, bred from *E. carnea* and *erigena*, is intermediate between its parents, reaching a height of 2 feet (60cm). Its cultivars sport a wide range of flower and foliage colors.

RHODODENDRONS: THE "ROSE-TREES"

The name of the genus known for its magnificent evergreen flowering shrubs is derived from the Greek words for "rose" (after the flower's general shape) and "tree." The 850 species are distributed in the wild throughout the northern hemisphere, although a significant number is concentrated in the mountains of Himalaya and China, with others in Indonesia and New Guinea. All demand lime-free soil and ample moisture during the growing season.

From the 1700s, a few species of rhododendron were introduced into western Europe from North America and western Asia. But the full potential of the genus was realized when botanist Joseph Hooker returned with his Sikkimese discoveries in the mid-1800s (see p.

discoveries were raised from the seeds he sent to Paris. Many were introduced to the West later by George Forrest, who worked in Yunnan between 1904 and 1931.

Delavay's colleague Jean Soulié (1858-1905) was yet another prolific French botanist-missionary. He was stationed in the isolated borderlands between Tibet and China, and became so fluent in the local Tibetan dialect that he was reputedly able to pass for a Tibetan. He collected and sent 7,000 specimens to Paris. Among them, and named in his honor, was *R. souliei*, a tall shrub with rounded leaves and flattened white or pale purple-pink flowers. Sadly, Soulié came to an unfortunate end. In 1904 a British military expedition to Lhasa stirred up trouble by ruthlessly attacking local people. Soulié was warned and did his best to escape with his specimens. But he was captured by the Tibetan authorities as a foreigner, tortured for fifteen days, and then shot.

Establishing hybrids

The majority of rhododendron species are found in the Himalayan-Chinese region. But a number of species used to develop the enormous range of hybrid rhododendrons found in gardens are natives of North America.

Until the introduction of Joseph Hooker's Sikkim species in the mid 1800s, the hardiest rhododendron known in Europe was *R. catawbiense*. This had arrived in 1809 from the Allegheny Mountains of southeast North America, where it is known as the mountain rose bay. It has broad, glossy green leaves and lilac-purple flowers. To combine hardiness with color, it was soon crossed with the similarly colored *R. ponticum* and the sulfur-yellow *R. caucasicum*, from the Pontus and Caucasus Mountains of northeast Turkey. This created the first of the hardy hybrids—large robust rhododendrons that thrive with little attention.

Later, exotically colored Asiatic species were crossed with *R. catawbiense* to produce still tougher plants with a wider range of colors. From the mid 1800s, hardy rhododendrons were planted widely in suitably lime-free garden soils across Europe and North America.

Azalea variations

Many deciduous rhododendrons were formerly included in the genus *Azalea*. Now they have been transferred to the genus *Rhododendron*, but the name of azalea has stuck for these and many small evergreen species which were introduced

to Europe from the early 1800s. To combine color, hardiness, long flowering season, and attractive scent, breeders began to cross the various American and Asian species with *R. luteum*, a yellow-flowered, strongly fragrant plant found from eastern Europe across to the Caucasus. Some of the first crosses were made in Belgium in the 1820s by P. Mortier, a baker by trade but also a keen amateur gardener. He used *R. luteum* as one parent, and *R. calendulaceum*, with yellow, orange, or red flowers from southern North America, or *R. nudiflorum*, a pink- or white-flowered species found throughout eastern North America, as the other. The result was the first of the now-

ANCESTRAL ROLE
Rhododendron luteum *the hardy yellow azalea from Europe and western Asia, was introduced to northern Europe in the 1790s and became an important ancestor of the Ghent azaleas bred in the 1820s. Completely deciduous, its foliage turns bright yellow and red before falling.*

famous 'Ghent' azaleas. These are fully hardy, funnel-flowered, and brilliantly colored.

Soon, several more North American species were added to the breeding programme, and the parentage of azalea hybrids became exceedingly complicated. Belgian and Dutch breeders crossed two Asian species, *R. molle* and *R. japonicum*, to create the 'Mollis' hybrids. These lack scent, but share the Ghent hybrid's vivid colors and late spring flowering.

Later, the Ghent and Mollis groups were crossed to create yet another series of complex hybrids. *R. occidentale*, from the mountain forests of California and Oregon, was also used in this program for its large, pink-flushed, fine-scented white flowers with a conspicuous yellow honey-guide patch. *R. occidentale* also flowers later than the other species, and this trait has passed to its hybrid progeny, thereby extending the azalea season into early summer.

Asian evergreens

In the wild, *R. simsii* is a small shrub from riverbanks of northern Burma, southern China, and the island of Taiwan. Although long cultivated in China, the first specimens to reach Europe in 1808 were the red-flowered wild plants. In later years, double-flowered and bicolored cultivars reached Europe from

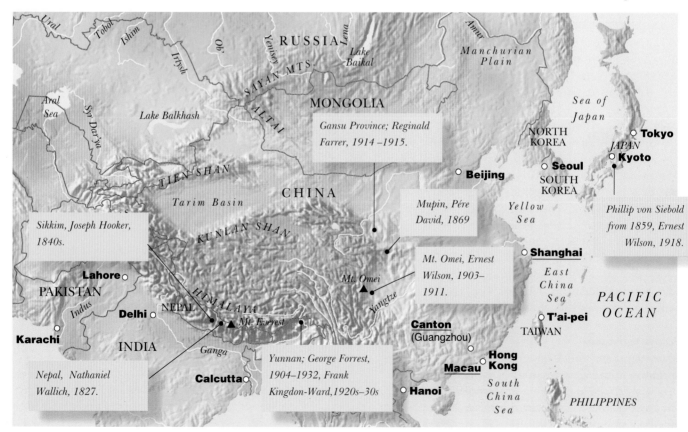

Gansu Province; Reginald Farrer, 1914–1915.

Mupin, Pére David, 1869

Phillip von Siebold from 1859, Ernest Wilson, 1918.

Sikkim, Joseph Hooker, 1840s.

Mt. Omei, Ernest Wilson, 1903–1911.

Nepal, Nathaniel Wallich, 1827.

Yunnan; George Forrest, 1904–1932, Frank Kingdon-Ward, 1920s–30s.

described by Wilson as the "foremost Japanese horticulturist of his time," took him to a nursery in Hatagaya district, not far from Tokyo. Wilson was bedazzled by the brilliant flowers of the Kurume azaleas, which then numbered some 250 cultivars. He asked the local experts to select six of the very finest for him. They included apple-blossom pink 'Takasago', 'Kurai-no-himo' with carmine-red flowers, and white 'Kure-no-yuki' with "hose-in-hose" flowers—a secondary petal tube or trumpet within the outer one. Wilson sent a consignment of 51 cultivars back to the Arnold Arboretum. These were nicknamed the "Wilson 50" and became the classic plants to propagate from.

THE STRAWBERRY TREE

Arbutus unedo is a small tree in the heather family, Ericaceae. It has a crooked trunk, smooth, bark, and its dense crown of evergreen foliage is bedecked in fall with flowers and fruits. It is found mainly in the warm conditions of Portugal, Spain, and other countries bordering the Mediterranean. Like many members of the family, its flowers are small, waxy, white bells. However this tree has two peculiarities: its fruits and its distribution.

The fruits, which develop over the course of a year into fleshy, bright-red berries have led to the common English name of "strawberry tree." The fruits look tempting, but they taste very insipid, as the species name *unedo* implies: "I eat one" (and only one!).

However, the fruit has other uses. In Portugal the berries are used to flavor a very potent liqueur known locally as *madrove*. When Spanish colonists explored the west coast of North America, now British Columbia, they encountered a similar species, *A. menziesii*, and named it the Madrona.

The strawberry tree has one oddly isolated population—in the groves of Killarney, in the far west of Ireland. The reason for this is probably the severing effect of glaciation combined with changing climate. *Arbutus* is a very ancient genus, shown by its distribution in both western North America and southern Europe. Populations have survived in areas too far south to be affected by the extensive northern glaciations of the Ice Ages. The Killarney trees cope because the local climate is exceptionally mild and moist. This is due to the tempering effects of the Gulf Stream, which brings warm water from the Gulf of Mexico to the northeast Atlantic.

Chinese gardens. These were not hardy enough for outdoor cultivation in European frosts—but indoors, they flowered in winter. For this reason the initially unpromising cultivars of *R. simsii* gained popularity as winter-bloom pot plants. They are now grown in enormous quantities in the Netherlands and Belgium for sale throughout Europe.

One of the most popular groups of dwarf rhododendrons is the Kurume azaleas. The open, upward-facing flowers of these low shrubs almost cover the foliage in a riot of bright colors. They were raised entirely in Japan, probably with a form of *R. kiusianum*, originally found on Mount Kirishima on Kyushu Island, as one parent. Their introduction to the gardens of North America and Europe is credited to English-born Ernest Wilson (1876-1930), one of the most successful of all planthunters in China and the Far East.

The classic "50" cultivars

Wilson undertook his first expeditions to China on behalf of the Veitch family who had been among the most important nursery owners in Britain since the early 1800s and responsible for introducing huge numbers of plants to general cultivation. Later, Wilson moved base to North America and was employed as a plant collector by the Arnold Arboretum of Boston, Massachusetts. Eventually, he became Director.

Between 1917 and 1918 Wilson made his last visit to the Far East, to Japan, on behalf of the Arnold Arboretum. His host Mr. Suzuki,

CREATED SPECIFICALLY TO
IMPRESS THE VISITOR WITH THE
MAGNIFICENCE OF THE FRENCH KING, THE GRAND, FORMAL DESIGN OF
VERSAILLES BECAME A MODEL FOR THE WESTERN WORLD.

THE GRAND FORMAL GARDEN

"The concept was undoubtedly that of the Sun King," Louis XIV (1638-1715). But the king wisely left the details of the architecture and garden design to Louis Le Vau (1612-70) and Andre Le Nôtre (1613-1700), whose earlier collaboration at Le Nôtre's first important garden Vaux-le-Vicomte was the prime influence in the creation of Versailles. Their Classical approach used the gardens as a setting for the architecture of the house. To achieve this Le Nôtre created wide open areas around the chateau, beyond which the trees and water features began. This space was filled with *parterres*—intricately-patterned flower beds—which stretch down to the "canal," a formal lake. Running east to west, the canal is aligned on the axis of the palace and provides the main vista. The sun appears to set at its far end. Many of the other features in the grounds also derived from the cult of the Sun King, especially the Classical statuary depicting the sun god, Apollo.

The canal is actually cross-shaped and was used by Louis for court entertainments in summer, with a special barge reserved for his musicians under the direction of Jean-Baptiste Lully (1632-87). Lakes and 1,400 fountains are an important feature of Versailles, but water was in short supply. To overcome this problem, massive engineering works brought water from distant sources to the estate, but even then there was not enough for all the fountains to play at once. As Louis strolled in the grounds, the *fontainier* controlled the fountains,

ensuring that those in view played, while those out of royal eyeshot were turned off. Louis spent a great deal of his time in the gardens, and wrote the "official guide" himself. There is a charming record of Le Nôtre's final visit to Versailles at the age of 85, when he and the King (a mere 62) were pulled around in special wheelchairs.

While Le Nôtre looked after the grand plan, the details were the work of Charles Le Brun (1619-90). As well as the ornaments in the parterres, Le Brun created the *bosquets*, ornamental groves that often surrounded features of statuary or architecture. To create an instant effect, mature trees were transplanted from the surrounding woods by a specially designed machine. Between the trees ran ornamental pathways, or *allées*, leading to distant corners of the 250 acre (101ha) park. At the far end of each *allée* was a feature to delight the visitor, ranging from the elaborate flowerbeds of the Grand Trianon, to the menagerie with its exotic animals.

GRAND SCALE
The vast scale, the grandeur, the overall pl[...] and the detail of Versaill[...] were intended to reflect [...] magnificence of the Sun King. The garden established the value of perspective in a planting scheme, of paths leading [...] special features, a Classi[...] statue, or a vistas, and t[...] extra dimension that wat[...] can give. There was also

horticultural
extravagance,
incorporating plants such
as orange and palm trees,
which were not naturally
suited to the climate. The
potted palms (above) have
to be overwintered in
greenhouses. Flowers were
brought from warmer
climates by the French
Navy and bedded out even
on the coldest days, to be
replaced if they died.

ARCHITECTURAL LINK
Formal parterres (left)
were made close to the
chateau to complement its
architectural
magnificence. Other
devices included hedged,
architectural style
enclosures (<u>cabinets de
verdure</u>)— and <u>bosquets,</u>
ornamental features
enclosed by trees.

$\mathcal{M}$ANY MEMBERS OF THESE
FAMILIES, WHICH RANGE FROM
FLAME-COLORED HIBISCUS AND PELARGONIUMS TO THE HUMBLE HARDY
GERANIUM, ARE HARDY PERENNIALS IN THEIR NATIVE LANDS BUT GROWN AS
ANNUALS IN COOLER REGIONS.

Zonal Pelargonium

$\mathcal{G}$ERANIUMS
AND $\mathcal{M}$ALLOWS
(GERANIACEAE AND MALVACEAE)

The three principal genera of the Geraniaceae family are named after long-billed wading birds, a tradition which started with the Greek physician Dioscorides in the first century A.D. He coined the name *geranion* from *geranos*—a crane—because the pointed fruit resembled a crane's beak. The Latin word *Geranium* was used for all members of the family

Pelargonium zonale, *an ancestor of the familiar red, pink, and white zonal pelargoniums of summer borders, was taken to* Holland from the Cape Province of South Africa in the late 1600s, together with the scarlet *P. inquinans;* they crossed, and the red "geranium" was born. If left unpruned, zonal pelargoniums become leggy shrubs like their wild ancestors.

THE FAMILIES' RANGE
across the world

Abutilon vitifolium, *a large shrub from the temperate mountains of Chile, is shortlived, but easily grown from* cuttings or seed. No *Abutilon* is totally hardy, but *A. vitifolium* will survive some frost, especially if it is grown against a wall for protection against the worst chills.

Tropic of Cancer

Equator

Tropic of Capricorn

South Atlantic Ocean

until 1789. Several 18th-century botanists noted that some geraniums were different from the rest, but none of them dared disagree with the great Swedish naturalist Carolus Linnaeus, who had decided all were bona fide members of the genus *Geranium*. The French botanist C-L. L'Héritier (1746-1800) rectified the situation in 1789, when he defined the genera *Pelargonium* (from *pelargos*, a stork) and *Erodium* (from *erodios*, a heron). Unfortunately, 200 years later, most pelargoniums are still called "geraniums" by the layperson. The difference between the genera is simple: a geranium flower has five petals of equal size, but a pelargonium has two which are different either in size or shape, and which are often differently colored or have a nectary groove.

Pelargonium, alias "geranium"

Before the colony of Cape Town was established at the Cape of Good Hope, at the southernmost tip of Africa in 1652, European trading ships sailing to and from India and the East stopped there for water. The Europeans' early colonization meant that many South African plants were known to the West long before those from other parts of the world. Pelargoniums, which are particularly abundant in the Cape region, were particularly sought. Although frost tender, most are relatively tolerant of drought and therefore easy to transport. The first species to reach Europe was *P. triste*—dull in color but sweetly scented—whose tuber acted as a storage system, enabling it to survive the journey without water. The English gardener, John Tradescant the Younger (1608-1662), received it in 1631 from his Parisian colleague René Morin (d. 1657?), a nurseryman who specialized in the importation of bulbs. Both believed that the plant originated in India, but all bar a handful of the 280 pelargonium species come from southern Africa. The remarkable perfume of *P. triste* is released at night but disappears during the day. It is still cultivated in glasshouse collections of succulent plants, or outdoors in frost free climates.

Robust reds

The red "geraniums" used as summer bedding plants are complex hybrids derived from *P. inquinans* and *P. zonale*, both of which were in

Hibiscus rosa-sinensis *is a dense shrub thought to be indigenous to southeast Asia, but its origins are blurred by its wide distribution in cultivation. There are* *cultivars with flowers in all shades of yellow, orange, pink, or red, many with a dark center.*

Pacific Ocean

Hibiscus schizopetalus *grows wild along the East African coast, where it may have been introduced from Asia in* *the distant past, but is now cultivated throughout the tropics. Forming a loose, open shrub up to 12 feet (4m) high, it should be given space to grow naturally without the need for pruning, which would reduce its elegance. It is too tender for outdoors in temperate regions.*

Key

	Tundra
	Boreal coniferous forest
	Temperate woodland and grassland
	Mediterranean
	Desert
	Tropical woodland and grassland
	Tropical rain forest

cultivation by the early 1700s. Seeds of *P. zonale* were sent to Holland from the Cape by the Governor, Wilhelm van der Stel, and plants were flowering in Amsterdam by 1700. It is a robust, scrambling plant, with white or pink flowers in the wild.

The scarlet coloring came from *P. inquinans*, an erect shrub from the eastern Cape. Hybrids between the two were highly sought by the mid 1700s, and reached their zenith in the bedding displays favored in the parks of northern Europe and North America during the mid to late 1800s. The most famous red pelargonium, whose intense, scarlet flowers have set a standard for all others since, was *P.* 'Paul Crampel', which was raised in France in 1903. Its breeder, Monsieur Crampel, realizing what a treasure he had, raised an enormous number of young plants, but to guard against theft and having his triumph preempted, prevented the plants from flowering until he was ready to release them onto the market.

Pelargoniums are too tender to survive outdoors in northern gardens, but because they are naturally acclimatized to long, dry, and sometimes cold periods in their homelands, they are relatively easy to overwinter in frost-free conditions. In mediterranean climates, pelargoniums can survive outdoors all year round, producing large, handsome shrubby plants which flower continuously.

Following the trail

The trailing, ivy-leaved pelargoniums of many a hanging basket and balcony, are derived from the scrambling South African species *P. peltatum*, while the regal pelargoniums with their large, ruffled, velvety flowers were bred from *P. cucullatum*. Both species are common in Cape Province, where they grow in sandy, well-

drained places in full sun—conditions which should be imitated in the garden. These, like many pelargoniums, are valued for their bold displays of flowers, but others, often with tiny flowers, are cultivated for their scented leaves. These are mostly selections, or primary hybrids, of wild species such as *P. tomentosum*, a spreading plant with large, lobed, and softly hairy leaves from shaded, moist places in the forests of the Cape—conditions it also favors in cultivation. Its foliage is strongly mint scented, and the essential oil which is the source of the smell has been extracted for use as a culinary flavoring. *P.* 'Graveolens', with its rose-scented, dissected leaves, and small, pale pink flowers, has also been grown commercially for its oil, especially on the island of Réunion in the Indian Ocean. Although it is not known in the wild, and is probably of hybrid origin, it has been cultivated in Europe since before 1792. A variegated sport, 'Lady Plymouth' has attractive grayish, white-bordered leaves.

The true geraniums

Cranesbills, or "hardy geraniums," are sometimes considered poor relatives to the gaudier pelargoniums, but many are first class hardy herbaceous plants which suit informal planting schemes. One of the best for herbaceous borders or naturalizing in wild gardens is *Geranium pratense*—the meadow cranesbill—which, in its wild form, is found in grasslands from western Europe eastward to China. In the Altai mountains of southern

Francis Masson reveals Cape diversity

Francis Masson (1741-1806), a Scot who became a gardener at the royal gardens at Kew in London, made his enduring mark on horticultural history in South Africa. He became the first in a long tradition of plant hunters from Kew when he went to South Africa to find new plants for the garden. Although he traveled widely in South Africa between 1786 and 1795, Masson based himself at the Cape, where he created a garden in which to grow his discoveries before sending them back to England.

At the time, the full diversity of the Cape flora was unknown, and almost everything Masson and his Swedish colleague Carl Peter Thunberg (1743-1828) found was new to western science. Among the hundreds of plants Masson sent to Kew were many pelargoniums, including *P. crispum*, a small shrub with citrus-scented foliage that is still popular. One plant of the cycad group, *Encephalartos longifolius*, collected by Masson himself during his first visit to the Cape (1772-1775), is still growing at Kew; it is the oldest pot plant in the world.

In 1779 Masson fought against the French on the island of Grenada in the Caribbean, and was taken prisoner; in the following year he lost all his possessions in a hurricane. His last trip to North America proved to be botanically unprofitable, although he is credited with introducing *Trillium grandiflorum* to Europe before his death in Montreal.

Siberia, swathes of *G. pratense* color the meadows a shimmering blue, but in Europe, intensive agriculture has often banished the species to the roadsides. It is a variable plant, and the flowers, which in the Altai are used for tea, may be white or pink instead of the usual blue. In Iceland a blue pigment was extracted to dye the garments of the fighting men—the Viking heroes of the sagas—from the 9th to 12th centuries A.D. Unlike the widespread *G. pratense*, *G. sanguineum* var. *striatum* (formerly var. *lancastriense*) is confined in the wild to sand dunes in a limited area of the Lancashire coast in northwest England. It is a form of the magenta-flowered "bloody cranesbill" but has paler, flesh pink flowers and prostrate stems. This highly localized, wild variation was in general cultivation by the early 1700s and became one of the few plants which the British Isles contributed to world horticulture. It grows well in well-drained soil.

MALLOWS: A TROPICAL ANCESTRY

Mallows share the five petals of Geranium flowers, but differ in the large number of stamens which are often fused into a tube around the style. The Malvaceae family is large, and its members, which include many woody species, are most abundant in the tropics. To add to the Pelargonium/Geranium confusion over names, a myth originating in the Mediterranean tells how the prophet Mohammed turned the humble mallow (*Malva*) into a geranium—although a Pelargonium is apparently meant(!)—in gratitude for a service it had done him. Versions vary, but one says the Prophet hung his shirt out to dry on the mallow, which became transformed by its contact with the holy object; another suggests that he was so delighted with some sheets made from mallow fiber that he upgraded the plant from which they came.

Plants for fiber

The cotton plant (*Gossypium* spp.) is among several members of the Malvaceae family which produce useful fibers to make cloth, although true mallows (*Malva* spp.) do not. The hollyhock (*Alcea rosea*), was once grown for its fiber, but was not very successful. It has been cultivated as a decorative plant in Europe since the Middle Ages; it is indigenous to the Near East and seeds were probably taken to northern and western Europe by soldiers returning from the Crusades. The soldiers added "holy" to *hoc*, the Anglo-Saxon word for a mallow.

The wild plants have pink or yellow flowers, but in cultivation many different shades soon appeared, including forms with deep, blood-red flowers which were so dark as to appear almost black. Another variety, with deeply incised, palmate leaves, became known in the late Middle Ages as *A. cannabina*, from its resemblance to the leaves of hemp, *Cannabis*

sativa, but it is now chastely known as *A. ficifolia*—the fig-leaved hollyhock.

Doubles and strains with variegated flowers were developed in European gardens, but disaster struck in the 1870s when hollyhock rust appeared. Although this disfiguring fungus seldom kills the plants, it stunts their growth, and hollyhocks never regained their former popularity. A smaller species from the Ukrainian steppes, *A. rugosa*, with large, pale yellow flowers, is less susceptible to rust, and makes an acceptable alternative to forms of *A. rosea*, especially where space is limited.

Tropical elegance

One of the most striking members of the mallow family is the hibiscus. In India, the flowers are used to polish shoes. Most cultivars grown throughout the tropics are derived from *Hibiscus rosa-sinensis*, which is believed to have originated in tropical Asia but was widely distributed around the Pacific long before the Europeans arrived there.

H. rosa-sinensis is the national flower of Malaysia, where it is known as "Bunga Raya;" the five petals are said to symbolize five National Principles: belief in God, loyalty to king and country, upholding the Constitution, the rule of law, and moral behavior. A closely related species, *H. schizopetalus*, is found along the East

African coast, and may represent a selection from *H. rosa-sinensis* carried there by early traders from India. It has elegant, pendulous red flowers with laciniate petals, and has become a common cultivated garden plant elsewhere in the tropics.

Although *H. rosa-sinensis* was first introduced to Europe in 1731 it requires warm greenhouse conditions to flourish there. More successful, as it is hardy throughout most of the north temperate region, is *H. syriacus*, which has been cultivated in western Europe for over 400 years. It is a stiff, erect shrub with the wide, funnel-shaped flowers typical of hibiscus in shades of white, red, and blue, which bloom over a long period in summer. Although it was first brought to the West from Syria—as its name suggests—it was in fact an ancient introduction to the Near East from India and China. The Chinese used the leaves to make tea, and ate the flowers.

Bell-shaped relatives

The name *Abutilon* comes from an Arabic word for a mallow. The abutilons are a genus of approximately 100 species from tropical and warm parts of the world; in tropical climates, to which they are best suited, they form large, vigorous shrubs. The best known are large-flowered hybrids with big yellow or red bells,

---

IN THE PATH OF THE CRUSADERS
When Crusaders of medieval times went to the Near East to defend the Christian faith against the threat of Islam, they picked up some tips on building fortifications from their enemies, and also some plants from the Mediterranean climates.

➡️ *main sea routes*

➡️ *main land routes*

(1) *First Crusade 1096-9*

(2) *Second Crusade 1147-9*

(3) *Third Crusade 1189-91*

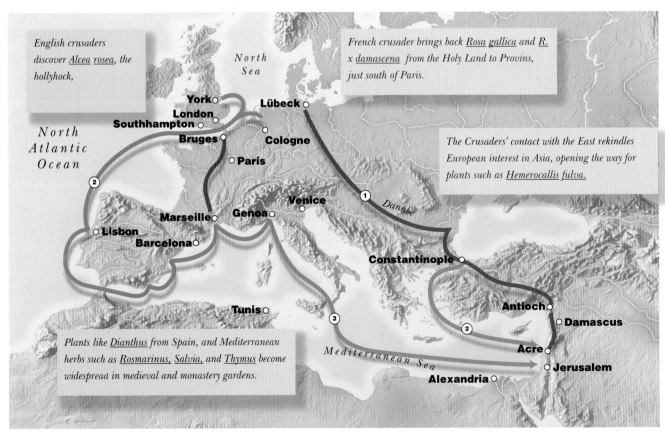

English crusaders discover <u>Alcea rosea</u>, the hollyhock,

French crusader brings back <u>Rosa gallica</u> and <u>R. x damascena</u> from the Holy Land to Provins, just south of Paris.

The Crusaders' contact with the East rekindles European interest in Asia, opening the way for plants such as <u>Hemerocallis fulva.</u>

Plants like <u>Dianthus</u> from Spain, and Mediterranean herbs such as <u>Rosmarinus</u>, <u>Salvia</u>, and <u>Thymus</u> become widespread in medieval and monastery gardens.

derived from the complex *A.* x *hybridum* These can be grown in greenhouses or for summer bedding in temperate regions, although a few are hardy enough to be grown outside in the mild parts. One is *A. megapotamicum* from Brazil, whose precise origin remains a mystery; its name, which means "big river," does not help pinpoint it in that land of great rivers.

A. megapotamicum is a scrambling plant, with narrow, triangular leaves and pendulous flowers, which needs the support of a trellis or another shrub. Yellow petals and chocolate brown stamens emerge from a dull crimson, inflated calyx beneath the leaves, giving the 1-2 inch (2.5-5cm) flower a tricolored appearance.

A. vitifolium, from Chile, is another erect shrub or small tree that is often short-lived in cultivation and is best renewed regularly from seed or cuttings. It has large, open, bell-shaped flowers of pale mauve or white, which, at the height of flowering, smother the plant. It, too, can be grown in sheltered situations in temperate climates.

DINNER PLATE SIZE
Breeders have created cultivars of <u>Hibiscus</u> <u>rosa-sinensis</u> (above) with dinner-plate sized blooms. Compact clones make good greenhouse plants in cold climates where the species cannot be grown outside.

CLIMATIC CHANGE
In warm climates, the remarkable tricolored flowers of <u>Abutilon</u> <u>megapotamicum</u> are produced in great abundance throughout the year; in colder areas they are seen only in summer.

*DELICATE IRISES OF ANTIQUITY,
DRIFTING CROCUSES, AND STATELY
GLADIOLUS WITH ORIGINS IN ARABIAN DESERTS AND MISSISSIPPI MARSHES,
SIBERIAN STEPPES AND MEDITERRANEAN SCRUBLAND.*

Iris 'Florentina'

THE IRIS FAMILY
(IRIDACEAE)

Iris was the Greek goddess of the rainbow, and her name is therefore an appropriate one for a genus containing more than 300 species of multicolored flowering plants which have been cherished for their beauty since antiquity. Irises are depicted on Minoan frescoes in Crete, and carved into the 3,500-year-old walls of the great temple of Karnak, at Luxor in Egypt. They were also popular subjects in the paintings of 16th and 17th century Dutch and Italian oil paintings, and resurfaced as popular subjects in the works of the Impressionists such as Claude Monet and Vincent Van Gogh.

There are several distinct genera within the Iridaceae family, ranging from the various groups of bearded, beardless, and bulbous irises, to crocuses, stately gladiolus, and finely scented freesias.

THE FAMILY RANGE
around the world

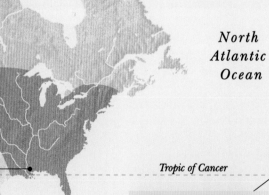

Arctic Circle

North Atlantic Ocean

Tropic of Cancer

Iris fulva has the reddest flowers of all wild irises, and has contributed red tones to hybrid offspring. Both "falls" and "standards" droop, giving the flower a characteristically floppy appearance. Its native habitat, in the marshes of the Mississippi valley in the southern U.S. is threatened by agricultural drainage or conversion of marshland to rice fields. Like many beardless irises, it is best grown in a rich, moisture-retentive, slightly acidic soil.

Iris pseudacorus, the yellow flag iris of European marshlands, is one of the finest bog and waterside plants for the garden, flowering prolifically in early summer. It is easily grown in moist soil or in containers submerged in a pond. The seeds float and are carried away from the parent plants by water; they were once recommended to be roasted as a coffee substitute.

IRISES WITH AND WITHOUT BEARDS

The bearded irises are so-named because of the tufts of pale hairs on the lower petals. They thrive in fertile soil and a sunny position, provided their rhizomes are planted level with the surface, and are not buried deeply.

Bearded irises are derived from the purple-flowered *Iris germanica* and its close relations, from Europe and the near East. They have been admired through the centuries for their large, wavy flowers, held high above the fans of gray foliage, and their hardy constitution. *I. germanica* was praised for its scent, as well as its flowers, in one of the earliest surviving pieces of horticultural writing. This was the poem *Hortulus* ("The Little Garden"), composed in about A.D.840 by Walafrid Strabo (c. A.D.809-849), Abbot of Reichenau, in Germany.

The origins of *I. germanica* are unclear. It may be an ancient hybrid which persisted because it had a particularly large share of so-called "hybrid vigor." It can survive years of neglect, and is almost impossible to kill, even by throwing it on a rubbish heap. Its many variants differ chiefly in their flower colors, most being

shades of purple. The distinctive white or yellowish beard of hairs occurs on the lower petals, which are known as the "falls." The upper petals, the "standards," lack the beard and may be paler in shade than the falls.

Bearded irises were developed in western Europe during the early 1800s, when various forms of *I. germanica* were crossed with two newly introduced species. These were *I. pallida*, a pale, lilac-blue species from the Balkans, and *I. variegata* from central and eastern Europe, a species with yellow standard petals and white falls intricately patterned with dark brown-purple lines.

These hardy bearded irises thrived in hot summers and cold winters, and generated a great deal of interest in continental Europe. The French nursery of Lemoine produced more than 100 distinct cultivars by 1840. Although most of these are no longer grown, they represented important stages or "building

Crocus chrysanthus, *and its hybrids with* <u>C. biflorus</u>, *like 'Blue Pearl' seen here, flower freely in early spring. Originating in Greece and Turkey, all need a warm, dry, summer rest period, but are tolerant of low winter temperatures, emerging and blooming as winter recedes. Available in many shades of white, yellow, blue, and purple, the hybrids can make an attractive mosaic when planted in a lawn or on the rock garden.*

Gladiolus murielae *is an elegant, sweetly scented species first discovered in Ethiopia, but widespread in the mountains of East Africa as far south as* Mozambique. *It is frost tender and best grown as a summer bedding or pot plant with the corms being lifted for storage in fall. Like many white-flowered plants, the scent is strongest in the evening when it attracts night flying moths that visit the flowers for their nectar.*

Key

Tundra

Boreal coniferous forest

Temperate woodland and grassland

Mediterranean

Desert

Tropical woodland and grassland

Tropical rain forest

blocks" in the development of modern bearded irises by nurseries and amateur growers in Europe and North America.

Orris and eternal wealth

One cultivar of *I. germanica*, 'Florentina', was once regarded as a distinct species. Its blooms have a delicious violet scent, while its dried, powdered roots are known as orris root and are used commercially to fix or stabilize the scent of pot pourri, in medicinal and skincare preparations, and in former times, for powdering wigs. "Orris" is a corruption of "iris" and for centuries Italy was a major producer.

A similar white iris, *I. albicans*, symbolizes wealth in Islamic tradition. Custom dictated its planting on the graves of Muslims, to ensure their prosperity in the next world. This practice ensured its spread as Islam expanded from its original lands to many countries in the Middle East and North Africa. Later the Spanish conquistadores took *I. albicans* to Mexico, where it thrived in the warmth, escaped cultivation and took to growing wild. It needs a warm spot in temperate countries, however, to do well.

Introducing different sizes

One group of irises was to have a dramatic effect on flower size. These were the tetraploid irises, introduced to Britain from the Middle East in the late 1899s. Tetraploid species have twice the usual number of chromosomes—the tiny packages of genetic material found in each of an organism's microscopic cells. This is a well-known natural phenomenon in many types of plants, usually leading to increased size.

Among that first batch of tetraploid irises was *I. cypriana*, which resembled *I. germanica*, but possessed flowers of almost twice the size—6 inches (15cm) in diameter. When crossed with existing cultivars, these tetraploids gave similarly large-flowered progeny. They were extensively used, especially in the U.S., to create the majority of bearded iris hybrids which are now familiar in temperate gardens throughout the world.

Breeders have also hybridized small-growing bearded species, such as *I. pumila* and *I. chamaeiris* from southern Europe, to create dwarf bearded irises.

The result is that today's bearded iris cultivars occur in a huge range of heights, from 6 inches (15cm) to as tall as a person. The color range is also enormous, from pure white through bright orange, deep blue to

near-black. But there is still no bright red bearded iris, although some cultivars achieve a glowing chestnut-red.

Beardless irises

In contrast to bearded irises, which like well-drained warmth, many beardless species are plants of wet or damp places. The common European species, *I. pseudacorus*, the yellow flag iris, is found wild in boggy places from Scandinavia east to Siberia, and south through Europe to North Africa. The variegated and pale-flowered garden forms of *I. pseudacorus* also flourish at the edges of ponds.

The flag iris is thought to be the inspiration for the floral symbol of France, the *fleur-de-lis*, which was adopted by King Louis VII in the 1140s as his emblem during the Second Crusade. The word *lis* may be a corruption of Louis, or it may refer to the River Lys, where the iris grows in profusion. It seems that Louis resurrected the floral symbol from a predecessor, Clovis I, King of the Franks who had adopted it as his family's emblem in the 6th century A.D. On a campaign in Germany, his army became hemmed in by the Goths against a bend of the River Rhine, near Cologne. Clovis noticed that at one spot, clumps of flag iris grew far out into the river and realized that the water must be shallow enough for his army to cross into safety, since the iris cannot tolerate depth.

North American and Eastern origins

Many species of beardless iris inhabit the marshes of eastern North America. However few are widely grown in gardens. *I. versicolor*

BEAUTY SUPPLEMENT
*Iris germanica 'Florentina'
(above) has been long
cultivated for its rhizomes,
which produce fragrant
orris powder used in
cosmetics.*

COLOR CHOICE
*While other iris cultivars
were selected for their
unusual flower shapes Iris
ensata (above) is prized in
Japan for its rich white,
blue, pink, or purple
flowers.*

RAINBOW COLORS
*A border of bearded irises
(right) portrays why the
genus is named for the
goddess of the rainbow.*

resembles *I. pseudacorus* in habit and preference for wet conditions. The flowers are pale purple, or rich red-purple in the variety 'Kermesina'. A species that has contributed to many hybrids is *I. fulva*, from the marshes of the Mississippi valley. Its rather flattened flowers are a rich chestnut-red, and it grows best in damp soil in warmer gardens. A handsome small species, *I. cristata*, grows at woodland edges.

In Japan, two damp-loving beardless irises were adopted as favored garden plants in the 6th and 7th centuries A.D. One was *I. laevigata*, found in marshes from Siberia to northeast Asia, and exceptionally hardy. In the wild it has dark violet (or occasionally white) flowers. This species is known in Japanese as *Kakitsubata*, a metaphor for "beautiful woman". Breeders developed numerous cultivars for the gardens of Japan's temples and nobility. In the oldest anthology of Japanese poetry, the *Manyoshu*, seven poems dating from about 760 A.D. praise the charms of *I. laevigata*'s violet flowers. In later centuries they featured in many works of art, including a screen known as *Kakitsubata-zu-byobo*, painted by Ogata Korin (1658-1716), and now classed as a Japanese national treasure.

The second beardless iris, *I. ensata* is another damp-meadow plant growing wild in Japan and also in Manchuria, eastern China, and adjacent areas of Russia and Korea. It resembles *I. laevigata* and is similarly hardy, but requires less water and so can tolerate moist, lime-free soil. *I. ensata* was cultivated in Japan for more than five centuries, producing a wide range of flower shapes and colors. It has recently become very popular in the U.S., where breeders have developed yet more cultivars, some with enormous double flowers.

Bulbous survivors

Many species of *Iris* from the warmer regions of southern Europe are bulbous, which enables them to become dormant and survive hot summers. One example is *I. xiphium*, the Spanish iris, a native of southwest Europe, from France to Portugal, and also North Africa. It was taken to northern Europe in the 1700s.

Iris xiphium was crossed with *I. tingitana*, a Moroccan species, to begin the line known as the "Dutch irises." Most were raised by the Netherland nurseries of van Tubergen in the early 1900s. They are a group of large-flowered, vigorous cultivars including the pale blue 'Wedgwood' and 'White Excelsior'. *I. latifolia*, is a similar species with larger, dark blue flowers, which is known as the "English iris." The name dates back to about 1571, when Flemish botanist Matthias de l'Obel observed the flower growing abundantly in gardens near Bristol, England. He assumed it to be native, but it is actually indigenous to the alpine meadows of the Pyrenees. It made its way via merchant ships from France and Spain to the port of Bristol, and still flourishes in the cool, moist climate of western Britain.

Winter delicacies

Although the small, bulbous, early-flowering species *I. reticulata* and its derivatives are frost hardy, they are often grown in pots in a cold greenhouse and brought indoors in early spring. However, even then the plants seldom survive and need annual replacement. *Iris reticulata* has red-purple flowers, and, like its close relation *I. histrioides*, which has larger, royal-blue flowers, is found in Turkey and the Near East. The two species have been hybridized to produce a selection of cultivars with flowers varying from bluish-white to rich red-purple.

Another winter-flowering iris, the rhizomatous *I. unguicularis*, produces scented, pale-lavender flowers in midwinter, when few other garden plants are in bloom. To perform well, it needs a perfectly drained position, in full sun throughout the year, conditions that echo its Mediterranean origins.

Temperate challenge

Many bulbous irises need hot, dry conditions and so present a real challenge to gardeners in temperate climates. They include the Oncocyclus iris species, which have large flowers mottled in muted shades of gray, purple, and brown. They have been prized in European gardens since the similarly colored *I. susiana*—the original mourning or widow's iris—was sent from Constantinople (now Istanbul) to Vienna in 1573. The flowers of *I. susiana* bear both a large black mark and a black beard at the base of the falls. The species name *susiana* is a probable corruption of the Arabic word for iris—*susan*. It is not easy to grow well, but many gardeners try as the plants are believed to be direct descendants of the original clone selected by the Ottoman Turks for export to Europe.

SMALL SCALE
Iris reticulata (above) flowers in early spring; it is one of several bulbous species from the Near East suitable for rock garden cultivation.

SAFFRON SPECIALITY
The spice saffron is produced from the scarlet stigmata of <u>Crocus</u> <u>sativus</u> (right), which must be individually handpicked, hence its high price.

CROCUSES FOR ALL SEASONS

Crocus vernus, which is common in northern Europe, is spring-flowering, but there are many species which bloom in the fall. One is *C. sativus*, which produces the spice saffron as a yellowish substance on the stigmas. *C. sativus* has one of the longest cultivated histories of any plant, for saffron has been an extremely expensive commodity for centuries, used in cooking, as a dyestuff, and as a medicinal ingredient. It is mentioned in the *Song of Solomon* from almost 3,000 years ago, and its cultivation is clearly portrayed in Minoan paintings dating to 1600 B.C. The name is of similar antiquity, being represented in ancient Sanskrit by the letters KRK, which ultimately became *Crocus* in Latin; the English "saffron" is derived from the Arabic *Za'ferán*.

It takes more than 4,000 flowers to produce just 1 ounce (28g) of dried saffron threads. Adulteration with inferior products has always been a problem, and since the Middle Ages, strict laws have been in force. At least one unfortunate trader in Germany was burnt to death in the same fire as his adulterated saffron. Cultivated saffron never grows as a wild plant. It is completely sterile, being propagated only from its corms. The first corms to reach England in the 1500s were smuggled, concealed in a hollowed-out pilgrim's staff. They helped to found a local industry in south-east England, around the town now known as Saffron Walden, Essex. Today, one of the major producers is Spain, where warm conditions encourage the crop. In more northerly regions the saffron crocus needs to be planted deeply in fertile soil, in full sun.

Drifting wild

The spring-flowering crocus, *C. vernus* is often planted in enormous drifts, in shades of purple and white, and is also popular as an early indoor flower. It is found in the wild across much of Europe, varying greatly in size and color. However it seldom reaches the dimensions of its now-familiar clones such as 'Remembrance' (purple), 'Pickwick' (striped) or the white 'Jeanne d'Arc'.

Crocus vernus is often planted with the large-flowered crocus 'Golden Yellow', another sterile plant that has been brightening gardens for more than 300 years—although purists insist these two should not be mixed. Other spring

THE ENGLISH GARDENING
TRADITION SCALED DOWN FROM THE
VAST LANDSCAPES OF THE 18TH CENTURY TO INTENSIVELY PLANTED
COTTAGE PLOTS RICH IN COLOR AND FRAGRANCE, WHICH BECAME AN IDEAL
FOR SMALL-SCALE GARDENING THROUGHOUT THE WORLD.

A Plantsman's Garden

Euphorbia characias and Erysimum

SWEET SHELTER
Borders close to the cottage are densely planted with choice plants which benefit from the sheltered position, together with traditional favorites such as forget-me-nots which self-sow.

Toward the middle of the 18th century, there was a reaction against the geometric formality of French garden style, to be replaced in England by the idealized landscape tradition. But this in turn, became obsolete in the early 1800s as many of the big estates broke up. Instead, plots of just a few acres close to the house were favored; wide open spaces were out, and full benefit was taken of the enormous variety of plants now available. The ultimate manifestation of this trend was the English cottage garden style, which has endured today as the way to make best use of small spaces, at least in temperate regions. One of the finest examples is the garden created by Bill and Joan Baker at their home, Old Rectory Cottage, in southern England. It demonstrates how the modern plantsman can draw inspiration from gardening history, and benefit from the efforts of centuries of plant collection and breeding.

Bill Baker himself follows in the tradition of great plantsmen of the past—from the Tradescants, John Bartram and Peter Collinson, to E.A. Bowles early this century—who took such pleasure in acquiring plants and passing them to their friends. Few visitors to Old Rectory Cottage go away without some growing memento of their visit. Not only is Bill a master plantsman who understands and appreciates the natural habitats and requirements of plants, but he has also collected many species himself on travels in North America, Europe, and Asia. Many plants that are now commonly grown in British gardens were bred in the garden of Old Rectory Cottage. The Bakers have even had cultivars named for them: a pink form of *Phlox carolina* is named 'Bill Baker', while his wife is honored by the pale lilac *Geranium phaeum* 'Joan Baker', which arose as a chance seedling in the

TRANSFORMATION
When Old Rectory Cottage was bought in 1957, it was surrounded by few old apple trees and some rather damp woodland, thick with weeds. Today, there is hardly a patch of bare soil visible: even the house walls are covered with plants, assimilating the house into the garden scene.

garden. Every part of the garden is occupied by plants, each one having its home carefully chosen according to its needs. A rock garden provides sites for alpine species, while the cottage walls are planted with climbers and tender shrubs. Parallel herbaceous borders full of choice plants focus a view through a gap in the hedge to a silvery poplar in the distance. Trees and shrubs were planted around the periphery, and despite its apparent informality, the orchard is equally carefully planned with rambling roses and clematis being encouraged to climb the apple trees. Sheets of spring bulbs are succeeded by drifts of *Geranium* species (one of the garden's specialties), and other herbaceous perennials.

COSMOPOLITAN

A tiny section of the garden—a border beneath a bay window—displays plants from all over the world.

1. Salvia grahamii
2. Crinum x powellii
3. Nerine bowdenii
4. Alstroemeria, Ligtu hybrids
5. Iris unguicularis
6. Abutilon vitifolium
7. climbing rose
8. Clematis
9. Galanthus reginae-olgae and Crocus speciosus
10. Cyclamen hederifolium
11. Liriope muscari
12. Lilium candidum
13. Galanthus reginae-olgae
14. Paeonia suffruticosa
15. Clematis

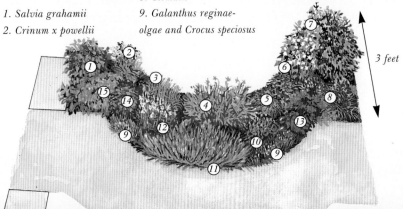

3 feet

flowerers, such as the *C. chrysanthus* hybrids or *C. tommasinianus*, may be planted in multicolored patches in well-drained conditions. They should increase in number to provide a fine display, as winter warms into spring.

GLADIOLI: LITTLE SWORDS

The name *Gladiolus*—'little sword' for the sword-shaped leaves—is given to a genus of 250 species that are popular as large hybrid cultivars, but which have yet to achieve their full potential as ornamental plants. Some 240 species are found in southern and tropical Africa, with the remainder in the Mediterranean

DRIFTS IN SPRING
Crocus vernus (left), the spring crocus from central Europe, often looks best when planted informally.

IMPORTANT PARENT
Gladiolus dalenii (below) is an important parent of large-flowered *Gladiolus* hybrids; it is found wild across much of Africa.

region. The only Mediterranean species widely grown in gardens is *G. communis* subsp. *byzantinus*, a magenta-flowered plant with a propensity to spread by seed.

The first African gladioli arrived in Europe in the 1700s, from the Cape of Good Hope. But their frost-tenderness made them difficult to grow until greenhouse technology advanced during the following century. One of the first successes was the cream-flowered *G. tristis*, which has a superb night scent and was in cultivation in the Chelsea Physic Garden, London, by 1745. In the early 1800s *G. tristis* was hybridized with the pink- or red-flowered *G. cardinalis*, also from South Africa, by London nurseryman James Colville. The resulting race of small-flowered hybrids, known as *G.* x *colvillei*—'Les Colvilles' in France—combined the scent of *G. tristis* with the color of *G. cardinalis*.

Toward the modern hybrids

The discovery of *G. dalenii* in Natal in the 1820s provided the foundation for the modern gladiolus hybrids. This species is so variable in the wild, throughout Africa and Madagascar, that botanists had classified collected specimens

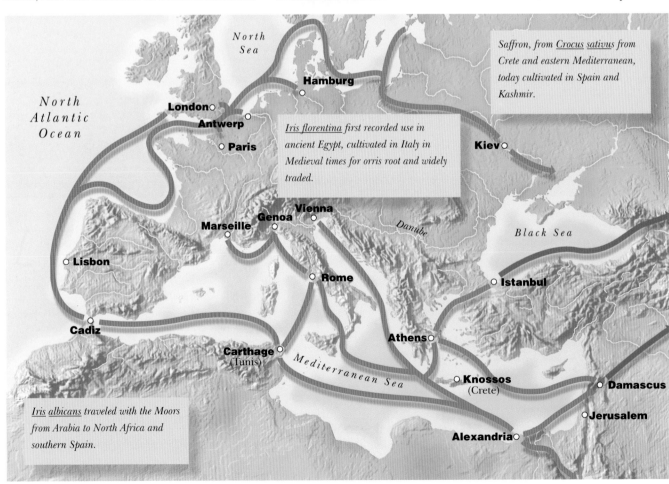

Saffron, from Crocus sativus from Crete and eastern Mediterranean, today cultivated in Spain and Kashmir.

Iris florentina first recorded use in ancient Egypt, cultivated in Italy in Medieval times for orris root and widely traded.

Iris albicans traveled with the Moors from Arabia to North Africa and southern Spain.

HIGHLY BRED
Large-flowered hybrid gladioli are the products of breeding work that has continued for nearly 200 years and incorporate genes from many different species.

CLASSICAL TRADING
Irises and crocuses were extra passengers on trading vessels linking the ports and cities of Europe with the northern Mediterranean and the Far East.

 Silk route

principal trade routes in Classical and Medieval Europe

with more than 30 different names—failing to realize that all represented the same species! In some areas the corms of *G. dalenii* are used for food, while in South Africa, it has medicinal applications for diarrhea and colds. The flowers are red, orange, or yellow, usually speckled all over with fine marks of orange or red, and the upper petal curves over the rest of the flower to create a hood. *G. dalenii* entered breeding work in 1837 with H. S. Bedinghaus, head gardener to the Duc d'Aremberg, at Enghien, Belgium. He crossed it with existing hybrids, most of which showed the influence of the open-faced flowers of *G. cardinalis*, and another red species, *G. cruentus*. The resulting hybrids took up the hooded flower characteristic of *G. dalenii*, but in a much reduced form, and were sold to the nursery of Van Houtte at Ghent. They were distributed under the name *G. x gandavensis*, from "Ghent" in Latin.

Variations in shape and color
French nurseryman Victor Lemoine (1823-1911) crossed *G. x gandavensis* with the hardy *G. papilio*, a somber species with deep purple blotches. Through further selection of the hybrids, his efforts yielded deep purple shades.

But the quality of hardiness was lost, leaving modern gladiolus hybrids susceptible to frost. The corms must be lifted and stored in a frost-free place over winter.

Lemoine also used *G. saundersii*, a red-and-white flowered species, and a green-flowered variant of *G. dalenii*, to expand the color range. A similar breeding programme by the German bulb-lover Max Leichtlin (1831-1910), of Baden-Baden, resulted in large-flowered plants with pale marks on the lower petals. Leichtlin's whole collection was sold to an American enthusiast, John Lewis Childs, in 1891. It has become the most important source of the modern 'Grandiflora' hybrids, mostly developed in North America. Their huge spikes of more than 25 large, ruffled flowers range from green to darkest purple.

The yellow gladiolus
Until 1904, no hybrid gladioli were yellow. This color was supplied when S. F. Townsend exhibited a yellow form of *G. dalenii*, then called *G. primulinus*, at a flower show in London. Townsend was chief engineer for the railway bridge that spans the Zambezi river at Victoria Falls in Africa. He had collected corms of a delicate, yellow-flowered gladiolus that grew within the misty spray of the falls. It became known as 'Maid of the Mist', and was soon being used by breeders to create a small-flowered race of cultivars known as the 'Primulinus' hybrids. These had hooded flowers, which were smaller than the 'Grandiflora' hybrids, but some also incorporated the yellow coloration in large-flowered cultivars.

Very few of the original wild gladiolus species are now grown in gardens or used in hybridization. An exception is *G. murieliae*, a new name for the plant formerly known as *Acidanthera bicolor*. Originating in Ethiopia, *G. murielae* was named after Muriel Erskine, whose husband collected corms and dried botanical specimens in 1931—although this species had already been subjected to a variety of botanical names stretching back to 1838. It is a tall plant, 4-5 feet (1.5m) in height, bearing an arching spike of white flowers blotched with maroon in their centers.

The blooms are strongly scented, especially in the evening, perhaps to attract their wild pollinators, moths. Muriel's gladiolus is not hardy in frost-prone areas, and so is usually grown as a summer bedding plant or greenhouse pot.

THE PODS AND FLOWERS OF THE
PEA FAMILY ARE AS DISTINCTIVE IN
CULTIVATED FORMS AS IN THEIR NATURAL HABITATS OF THE AUSTRALIAN
AND AFRICAN BUSH, THE HEATHLANDS AND MOUNTAINS OF EUROPE AND
ASIA, AND STEAMY SOUTH AMERICAN RAINFORESTS.

Russell lupin

THE PEA FAMILY

(LEGUMINOSAE)

Members of the pea family, such as various forms of pea, peanuts, and beans, are, after grasses and cereals, the world's most important source of agricultural foodstuffs.

However the legumes, as they are known, also embrace a large number of flowers, bushes, and trees which have become familiar ornamental plants. The cultivated legumes, which have made often lengthy journeys from natural habitats all over the world, range from sweetly scented "sweet peas" and laburnums native to Europe, to American lupins, Chinese

Lupinus polyphyllus *comes from the shingly edges of rivers in western North America, where planthunter David Douglas noted that the usual blue-flowered plants occasionally gave rise to white-flowered variants. It became an important ancestor of the cultivated garden lupins developed by in the 1900s.*

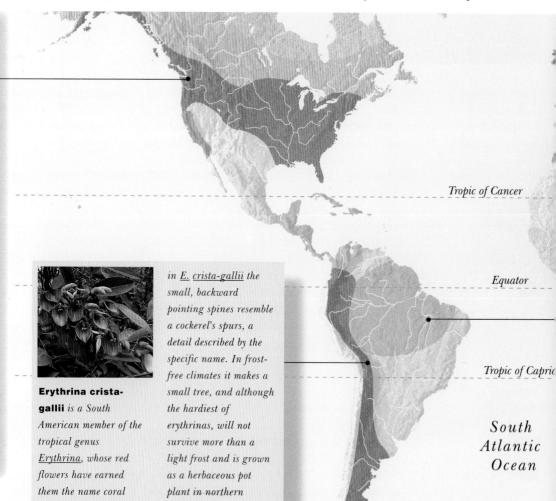

Erythrina crista-gallii *is a South American member of the tropical genus* <u>Erythrina</u>*, whose red flowers have earned them the name coral trees. Many are spiny;*

in <u>E. crista-gallii</u> *the small, backward pointing spines resemble a cockerel's spurs, a detail described by the specific name. In frost-free climates it makes a small tree, and although the hardiest of erythrinas, will not survive more than a light frost and is grown as a herbaceous pot plant in northern gardens.*

Tropic of Cancer

Equator

Tropic of Capric

South Atlantic Ocean

wisterias, and acacias and wattles from Africa and Australia.

The botanical name of the pea-and-bean family comes from the Latin word *legumen*, meaning "pea pod." It refers to the characteristic structure of the fruit with its two

long sides or valves, which split to reveal a row of seeds. All members of the family have pods, although in some species they are woody rather than fleshy.

How the pea became "sweet"

The garden pea, *Pisum sativum*, provided the key to unlocking the secrets of heredity and plant breeding (see p.102-3). A close relation which has been greatly modified by diligent breeding work is the sweet pea, *Lathyrus odoratus*, whose fragrant flowers in shades of white, pink, red, and purple are descended from a much less showy ancestor—the wild sweet pea, a native of Sicily and southern Italy.

This wild ancestor is an easily grown, though rather straggling annual whose stems bear two small flowers, dark bluish-purple and maroon-red, with outstandingly sweet fragrance. It was first described for science in 1697 by Father Francisco Cupani (1657-1711), a Sicilian priest. In 1699 he sent seeds from Sicily to England, to his friend Dr. Robert

Ulex europaeus *is the gorse or furze of European heathlands, where it usually grows on acidic soil. A densely prickly shrub, it covers itself each spring in golden flowers that smell of coconut. Valued for its ability to stabilize sanddunes, gorse has been taken to many parts of the world, where it has sometimes become a nuisance.*

Key

	Tundra
	Boreal coniferous forest
	Temperate woodland and grassland
	Mediterranean
	Desert
	Tropical woodland and grassland
	Tropical rain forest

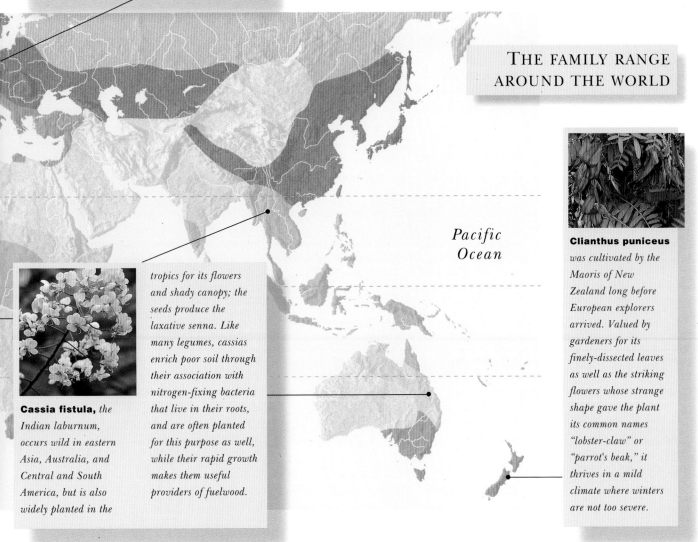

THE FAMILY RANGE AROUND THE WORLD

Pacific Ocean

Cassia fistula, *the Indian laburnum, occurs wild in eastern Asia, Australia, and Central and South America, but is also widely planted in the tropics for its flowers and shady canopy; the seeds produce the laxative senna. Like many legumes, cassias enrich poor soil through their association with nitrogen-fixing bacteria that live in their roots, and are often planted for this purpose as well, while their rapid growth makes them useful providers of fuelwood.*

Clianthus puniceus *was cultivated by the Maoris of New Zealand long before European explorers arrived. Valued by gardeners for its finely-dissected leaves as well as the striking flowers whose strange shape gave the plant its common names "lobster-claw" or "parrot's beak," it thrives in a mild climate where winters are not too severe.*

SWEET SCENT OF PEA
modern cultivars of
Lathyrus odoratus (above)
often lack the scent that
made their ancestors so
distinctive and gave them
the name "sweet."

BORDER MAINSTAYS
Russell lupins (far right)
are the result of decades of
patient selection from
many different North
American species.

Uvedale, headmaster of Enfield Grammar School, near London, a leading gardener of his time. From these early blue and red plants, a white form arose a few years later, as frequently happens when many plants are raised from seed. In 1731 a pink form appeared and became known as 'Painted Lady'. It was long rumored that this form had originated in Ceylon (now Sri Lanka), but sweet peas are unknown there. 'Painted Lady' is still grown, and its origins remain a mystery.

Breakthrough developments
The early cultivated sweet peas were valued primarily for their strong scent. But, as with many other plants, their development was slow because growers relied on natural pollination to increase the range of colors. From the mid 1800s breeders learned to cross-pollinate flowers, so that individual desirable features could be combined.

Henry Eckford (1823-1905) of Cheshire, England, began breeding sweet peas in 1870, when there were only 15 cultivars. By 1900 he had personally raised over 100 new and distinct forms. These had much larger flowers, ranging in color from cream to deep burgundy, and with four blooms per stem instead of two.

An even greater breakthrough came in 1901 from Silas Cole, gardener to Earl Spencer of Althorp in Northamptonshire, England. He

exhibited a sweet pea with frilled petals, named 'Lady Spencer' in honor of his employer's wife. The Spencer name became attached to all frilled sweet peas, and they rapidly almost replaced the plainer, old-fashioned varieties, at least for a time.

For most of this century, sweet pea breeding has been dominated by the Unwins, a British nursery family which is still raising new cultivars. A typical example of the modern Spencer sweet pea is 'Diana' after Diana, Princess of Wales (1961-1997), formerly Lady Diana Spencer of Althorp. Its large, glowing orange-red flowers have a pronounced frill.

As Spencer-type sweet pea flowers became larger and frillier, the scent of old varieties which earned them the descriptive of "sweet" diminished, with the consequence that many modern cultivars lack fragrance. Breeders are now trying to develop flowers that combine both large flowers and fragrance. But some gardeners still prefer the older, strongly-scented cultivars such as 'Painted Lady' and the purple-and-blue form 'Matucana'. which closely resembles the wild species.

Wild American lupins
In 1825, Scottish planthunter David Douglas explored the vegetation of what is now Oregon and Washington State, U.S. Among the many seeds he sent back to his employers, London's Royal Horticultural Society, were no fewer than 23 species belonging to the legume genus, *Lupinus*. One in particular, *L. polyphyllus*, was a handsome, herbaceous perennial reaching 5 feet (1.5m) or more, with long spikes of blue, or occasionally white, flowers. Douglas found it growing in enormous numbers along shingle riverbanks, and it soon became almost as common in European gardens.

Lupins are also known as "lupines" in North America. Indeed this is predominantly an American genus, with only a few members from Europe and Africa. Most species grow in open, dry places, from the southern Andes north to Canada. The genus name of *Lupinus* is often thought to derive from the Latin word *lupus*, meaning "wolf." In fact it comes from *lupe*, a Greek word meaning "grief." This may be a reference to the seed's bitter flavor and poisonous content.

Lupins also had a reputation for exhausting their soil—a voracity that enhanced the wolfish myth throughout Europe. In Dutch, German, and Swedish, the common names

translate as "wolf's bean." But the opposite applies. Like many legumes, lupins enhance soil fertility because they harbor beneficial, "nitrogen-fixing" bacteria in their roots, often in lump-like nodules. These microbes extract nitrogen gas from the air and help to convert it into nitrate mineral salts in the soil. This is the natural equivalent of adding artificial nitrogen-rich fertilizers to the soil.

Cultivated success

There are some 200 species of Lupinus, but apart from *L. polyphyllus* and the Russell lupins described below, relatively few are commonly cultivated. Annual species such as *L. texensis* and *L. subcarnosus*, known as "the blue bonnets of Texas," sometimes decorate wild flower gardens with their bright blue flowers.

The flowers were planted out in many European gardens, but did not attract the attention of plant breeders until 1911. An English gardener, George Russell of York (1857-1951), decided, at the age of 60, to take the matter in hand. On his allotment he gathered together lupins of many species, including the Mexican annual *L. hartwegii* with pale blue flowers, and varieties of the abundant *L. polyphyllus*. Then he let the bees go to work. Each year he saved and sowed the seeds, ruthlessly weeding out any plant that did not improve on its parent.

After 15 years, Russell began to obtain plants approaching his goal. But it was not until 1937 that he exhibited his new varieties, now in shades of pure red, pink, orange, and yellow, and with bicolored flowers in blue and yellow, purple and yellow, and pink and amethyst. As an amateur gardener, and using the simplest of methods, George had created an entirely new category of garden plant. His creations are known as "Russell lupins," and a variety with creamy-pink flowers is named 'George Russell'. One shortcoming of Russell lupins is that, with annual species among their parents, they are shortlived and so must be renewed regularly from cuttings or seeds.

Wisterias from China

A cardinal rule of plant science states that the spelling of a botanical name, once agreed and published, cannot be changed—even if it contains a mistake. Otherwise one genus of vine-like climbing legumes would be known as "wistarias." The name honors Caspar Wistar (1761-1818), Professor of Anatomy at the

University of Pennsylvania, but a spelling mistake has immortalized these plants as wisterias. *Wisteria sinensis*, as the specific name suggests, is a Chinese plant. It was introduced to European gardens from the nurseries of Canton, in southern China, by John Reeves. The first two plants were entrusted to two different ship captains, who arrived in England within a few days of each other, in May 1816. They were Captain Robert Wellbank and Captain Richard Rawes of *Camellia* fame (see *Camellias and Magnolias* chapter).

Wisteria has the knack of looking as if it has always been growing on any ancient building, even if the plant is only a decade or two old. It is a feature of late spring in temperate gardens around the world, with the usually pale, blue-purple flowers appearing shortly before the leaves.

Mount Fuji connections
Wisteria sinensis is distinguished from its Japanese counterpart, *W. floribunda*, by its shorter inflorescences, fewer leaflets, and a stem that twines in an anticlockwise direction. *W. floribunda* twists clockwise as it grows, while their hybrids do not know which way to go! *W. floribunda* is extremely vigorous, with stems 60 feet (18m) or longer, and looks particularly fine when permitted to grow into a tree, from which its flowers hang elegantly.

Wisteria floribunda is known in Japan as fuji. However this word also means "wisteria-color."

ESTABLISHED CHARM
The pendulous spring flowers of Wisteria *add a sense of aging elegance to buildings young and old. Although usually mauve, white and pink cultivars are also grown.*

WELL TRAINED
Laburnum x watereri *is a small tree which can be easily trained to form a tunnel (far right). Here it contrasts strikingly with the round purple heads of the alliums.*

So the volcanic cone of Fujiyama (Mount Fuji) may be translated as "wisteria mountain," or "wisteria-colored mountain." *W. floribunda* has been popular in Japan for centuries. Typically its flowers are silvery-lilac but many darker cultivars have been named. Kokkuryu Fuji or 'Black Dragon' has large, deep-purple blooms; a white-flowered clone with long inflorescences is named Shiro Naga Fuji or 'White Snake'.

Jade vines and laburnums
The jade vine, *Strongylodon macrobotrys*, is a wisteria-like climber for tropical gardens, or greenhouses in temperate regions. This native of the island of Luzon, in the Philippines has flowers in luminous jade green, with touches of aquamarine, shaped rather like a lobster's claw.

A similar flower shape, but colored bright red, belongs to *Clianthus puniceus*, a shrub from New Zealand. It is non-climbing in habit, but is often trained against a wall in temperate gardens to benefit from a protected position. The flower shape and color have led to common names such as "lobster's claw" and "parrot's beak." Ship's botanist Joseph Banks (see *The Heather Family* chapter) visited New Zealand on Captain James Cook's first round-the-world voyage, in 1769-70. He found *C. puniceus* growing as an ornamental shrub by the houses of the native Maori people. Wild specimens were discovered 55 years later, growing in a few places on the North Island coast. Some wild specimens may be more than a hundred years old, but in cultivation the plant tends to be short-lived.

The pendulous, bright yellow flowers of the laburnum or "golden rain tree," are a common sight in temperate gardens in late spring. Laburnums are naturally small trees which can be trained to form an arch or tunnel—a spectacular sight in full flower. Most commonly grown is *Laburnum* x *watereri*, a hybrid between two central European species, *L. alpinum* and *L. anagyroides*, which was first recognized in an English nursery during the 1800s, but has since been found in the wild. The clone 'Vossii' freely produces inflorescences up to 18 inches (45cm) long.

Acacias and wattles
The legume genus *Acacia* includes the acacias and thorn trees of Africa and the wattles of Australia—more than 1,200 species in total. They are evocative symbols of savannah, bush, and outback. In Africa, as the name suggests,

most species are savagely thorny as defence against browsing mammals. In Australia, where mammals are less threatening, wattle trees lack such weapons.

Wattles are tremendously diverse, but a typical species is *Acacia dealbata*, the silver wattle. This small tree, wild in southeast Australia, which grows to 30 feet (about 10m) high, with feathery, bluish-gray leaves, and abundant, gold-yellow flowers in late winter, is one of the hardiest species, surviving introduction to Europe in 1820. It thrives in warm temperate lands, and has widely naturalized in southern Europe. Farther north it survives only where temperatures seldom drop below freezing. Cut stems are sold in florist's stores as "mimosa."

The fever tree

The yellow-barked fever tree, *Acacia xanthophloea*, is an African species which is widely cultivated for its ornamental bark and ability to withstand occasional flooding. Rudyard Kipling, in his *Just So Stories*, described it growing along the banks of the "great, gray,

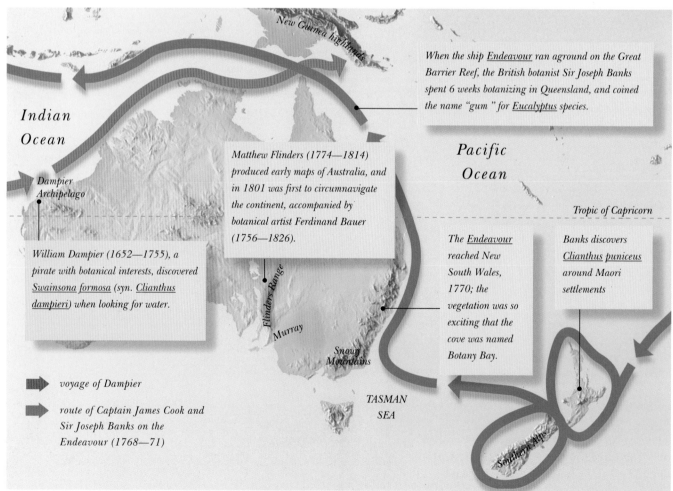

When the ship Endeavour ran aground on the Great Barrier Reef, the British botanist Sir Joseph Banks spent 6 weeks botanizing in Queensland, and coined the name "gum" for Eucalyptus species.

Matthew Flinders (1774—1814) produced early maps of Australia, and in 1801 was first to circumnavigate the continent, accompanied by botanical artist Ferdinand Bauer (1756—1826).

William Dampier (1652—1755), a pirate with botanical interests, discovered Swainsona formosa (syn. Clianthus dampieri) when looking for water.

The Endeavour reached New South Wales, 1770; the vegetation was so exciting that the cove was named Botany Bay.

Banks discovers Clianthus puniceus around Maori settlements

New Guinea highlands

Indian
Ocean

Dampier
Archipelago

Pacific
Ocean

Tropic of Capricorn

Flinders Range

Murray

Snowy
Mountains

TASMAN
SEA

Southern Alps

➡ *voyage of Dampier*

➡ *route of Captain James Cook and Sir Joseph Banks on the Endeavour (1768—71)*

green, greasy Limpopo River." Its flowers are less spectacular than those of other acacias, but the soft greenish-yellow bark is extremely attractive, especially in the rays of a setting sun. This is a plant for warm gardens in the tropics and subtropics, since it is native to eastern and southern Africa, and is frost-sensitive. African explorers believed that it gave out vapors which caused malaria, hence part of the common name. In fact, malaria is spread by the mosquitos that breed in the swampy water of its natural habitat.

The sensitive mimosas

The sensitive plant, *Mimosa pudica*, was introduced to Europe in 1637 by the English gardener John Tradescant the Younger. Its "moving leaves" immediately fascinated the gardening public. In fact they are leaflets, many making up one leaf. When touched or shaken, the leaflets fold up like the curling fingers of a hand, and the whole leaf droops downward. Among plants, this almost animal-like movement—taking just a second—is exceeded in speed only by the Venus fly trap.

A native of South America, and naturalized in many tropical countries, this mimosa needs to be grown as a potplant in cooler regions, usually as an annual. It bears small, round heads of pink flowers which, like all acacias and mimosas, have numerous long stamens forming a fluffy mass.

The mimosa's movement has been used for various purposes through history. In 1739, French Jesuit missionary Pierre Nicholas Cheron d'Incarville (1706-1757) was sent to China, hopefully to convert Emperor Chien Lung to Christianity. D'Incarville had heard that the Emperor was curious about natural history, so prepared himself by undertaking tuition in botany. But on his arrival at the Imperial Court in Beijing (Peking), he was given the post of master glassmaker. Eventually, however, he was able to present the Emperor with two sensitive mimosa plants. Their antics fascinated Chien Lung so much that d'Incarville won his confidence. He was allowed to visit the Imperial Gardens and send seed and plants home to France.

Flaming Fancy-Anna

The English names of "flamboyant," "peacock," or "flame tree" salute the brilliant scarlet flowers of *Delonix regia*. This woody member of the Caesalpinioideae subfamily of legumes is

one of the most widely planted garden and street trees in the tropics, valued for the shade provided by its leafy, flat-topped crown as well as its flaming floral display. The blooms develop into pods that ripen over a period of two years and may be 2 feet (60cm) long. When the sun's heat finally warps the woody pod-walls, they split to reveal bean-like seeds set into individual woody cavities.

The flame tree was discovered for science in 1828 by Austrian botanist Wenzel Bojer (1797-1856), in a stand of trees probably planted by Arab settlers, on the east coast of its native Madagascar. Bojer collected the seeds which have yielded all the generations of planted trees across the tropics. Indeed, wild specimens were not discovered until 1932, on limestone cliffs in western Madagascar.

This species has innumerable local names across the tropical world. In Jamaica it is "fancy-anna," a corruption of *Poinciana*, a related genus to which the tree was first assigned, before being renamed as *Delonix*.

THE JUDAS TREE *Cercis siliquastrum* (below) acquires its common name from the legend that the disciple of Jesus Christ, Judas Iscariot, hanged himself from its spreading boughs.

From peas to plant breeding

The garden pea, *Pisum sativum*, is seldom grown as an ornamental plant but it has a hallowed place in the annals of plant history. This was the species bred by Gregor Mendel (1822-1884), a monk from Brno (now in the Czech Republic). His experiments in the Abbey gardens in the 1850s helped to unlock the secrets of heredity and genetics, and put plant breeding onto a scientific basis for the first time. Gardeners and plant breeders still use some of his methods today, especially selection and hybridization, to develop a new cultivar.

Parental influence

All members of a species vary from each other, because they have inherited different selections of genes from their parents (apart, that is, from clones propagated by cuttings or other asexual means). The gene is the "unit" of inheritance and controls some feature or trait of an individual. This feature may be an obvious physical characteristic, such as blue flowers in the masses of lupins, *Lupinus polyphyllus*, observed by David Douglas in northwest North America. A different version of the flower color gene is responsible for the occasional white flowers. Mendel studied the pattern of inheritance of the height trait—tall and short forms of garden peas.

Genes also control less obvious features or traits, such as resistance to disease, greater tolerance of cold conditions, or ability to utilize certain soil minerals. Almost all aspects of life are governed or influenced by inherited genes. As life continues and evolves, generation after generation, certain features come and go in natural populations. If a certain feature helps the plant (or animal) to survive, it will probably spread, over many generations, to other members of the population. This can be viewed as nature favoring or selecting the feature.

Gardeners can take the place of nature by choosing individual plants with desired features and use them to produce the next generation, while discarding the rest. This is a manmade process—artificial selection, as opposed to natural selection. For example, in a field of lupins, the rare white-flowered individuals could be kept, while the blue ones are thrown away. Seeds sown from these white survivors are then subjected to the same

processes, and so on. Over generations, a distinct cultivar—where the desired characteristic or trait is more or less "fixed", occurring in most or all individuals—can be developed. The cultivar name is always indicated in single quotation marks, such as *Lathyrus odoratus* 'Diana'.

Choice of parents

In nature, the chance processes of pollination by bees, birds, and other creatures determine which individuals of the same species breed together. Hybrids, or crossbreeds between different forms or species, are usually rare in

SLOW DEVELOPER
Delonix regia is a native of Madagascar that has been planted throughout the tropics. The brilliant scarlet and white flowers are followed, as in all legumes, by a pod-shaped fruit, although this may take two years to open.

the wild. However, in outdoor and greenhouse cultivation, the plant breeder has considerably more control. The pollen, containing the male sex cells, can be transferred by hand from what is effectively the father plant to the mother plant. This process of crossing or "choosing your parents" allows the gardener to bring together genes in specific and desired combinations. The resulting seeds are sown, the desirable offspring are selected, and the process of hybridization between chosen parents is repeated.

Geneticists refer to the first generation from any cross as the F1 generation, but gardeners usually use it to describe the hybrid—whose offspring, the F1 generation—will be uniform in quality. Such uniformity may be an important consideration if a massed display is required. F1 seed usually costs more than ordinary, open-pollinated seed because the production costs required to produce these controlled hybrids are much greater than for seeds produced in an open field. However, seeds raised from expensive F1 plants will not prove as satisfactory, since genetic variation is expressed in the second generation, the F2. If F1 quality plants are required, seed must be repurchased annually.

*R*EPUTATIONS, FAME AND
FORTUNES HAVE BEEN MADE AND
LOST AS LILIES, TULIPS, AND FRITILLARIES HAVE JOURNEYED AROUND THE
WORLD FROM THE WILD TO CULTIVATED SERENITY.

Lilium candidum

THE LILY FAMILY
(LILIACEAE)

The white lily, *Lilium candidum*, was possibly the first plant in history to be specially grown for its flowers. In the eastern Mediterranean, it is depicted on Minoan frescoes in Crete dating back more than 3,500 years, and on Assyrian bas-reliefs some 2,700 years old. From the first centuries A.D., the glistening white, richly scented flower was adopted as a symbol of purity by the Christian Church and became associated with the Virgin Mary. However the name "Madonna lily" was not coined until the 1800s.

In Ancient Rome the scaly bulb of *L. candidum* was considered medicinal, and ground to form poultices for wounds and sores. Roman legions are reputed to have grown it near roadsides for the benefit of footweary soldiers. And it is indeed found growing wild in many

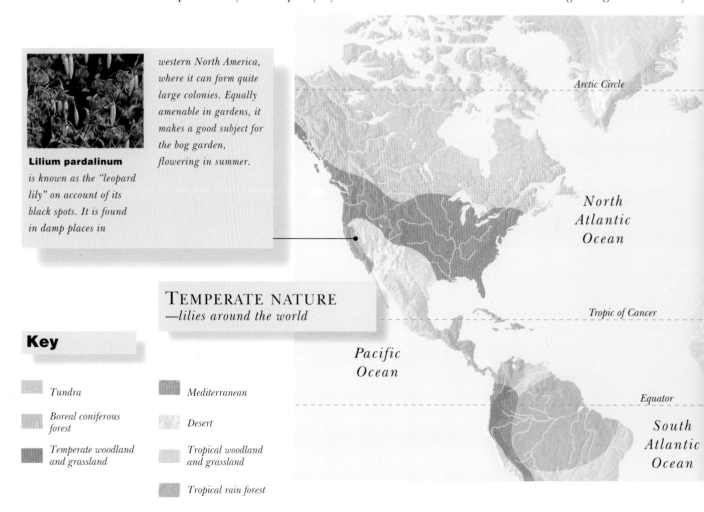

Lilium pardalinum

is known as the "leopard lily" on account of its black spots. It is found in damp places in western North America, *where it can form quite large colonies. Equally amenable in gardens, it makes a good subject for the bog garden, flowering in summer.*

TEMPERATE NATURE
—lilies around the world

Arctic Circle

North Atlantic Ocean

Tropic of Cancer

Pacific Ocean

Equator

South Atlantic Ocean

Key

Tundra	Mediterranean
Boreal coniferous forest	Desert
Temperate woodland and grassland	Tropical woodland and grassland
	Tropical rain forest

parts of the former Roman Empire. However, away from its wild home of the warm Balkan hillsides, it can be difficult to cultivate. It requires warm, well-drained conditions with plenty of sunlight and air or it may develop the debilitating fungal disease, botrytis.

Easier substitutes

The white Easter lily, *L. longiflorum*, originally from the Ryukyu Islands of southern Japan, although rather frost tender, is easier to grow. Its long, yet wide-mouthed trumpet flowers

Fritillaria meleagris
is a native of damp meadows in northern Europe, where it sometimes grows in millions, coloring the

fields purple. White-flowered forms often occur. Other European species have fascinating, if somberly colored flowers, in shades of green, brown, and purple, but these tend to be less easy to grow successfully.

earned it the Japanese name *Teppo-yuri* meaning "blunderbuss lily". A fragrant Chinese species, the "regal lily," *L. regale*, is exceptionally easy to grow from seed in fertile soil, in sun or light shade. On seeing it in the wild for the first time, its discoverer Ernest Wilson wrote: "it is crowned with large, funnel-shaped flowers, more or less wine-colored without, pure white and lustrous on the face, clear canary yellow within the tube, and each stamen tipped with a golden anther." *L. regale* flowers in the third season after sowing from seed; it is worth taking precautions to guard against late frosts which may damage young shoots.

A hair-raising adventure

Ernest Wilson (1876-1930) discovered the regal lily growing in "hundreds, in thousands, aye in tens of thousands" on the hills of west China's Sichuan Province in 1903. In 1904, he sent 300 bulbs to his employers, the English nursery of James Veitch and Sons. In 1910, he collected more than 6,000 further bulbs, to send to America. But on his return, a landslip swept his sedan chair hundreds of feet into the valley

Colchicum speciosum *'Album' is considered by many gardeners to be the most beautiful of fall-flowering bulbs. The leaves emerge in the following spring. C. speciosum is usually pink, and occurs in the subalpine meadows of the* *Pontus and Caucasus mountains. Other species occur in similar places throughout Europe and central Asia. All colchicums are poisonous and the large leaves are avoided by grazing animals.*

Indian Ocean

Tulipa clusiana, *the "lady tulip," is named for Carolus Clusius, the unwitting founder of the Dutch bulb industry, who introduced it to Europe in the late 1500s. It seems to be an old garden selection from the wild T. clusiana ssp. stellata, which grows on* *the dry hillsides of central Asia as far east as the Hindu Kush and Pakistan. In gardens it does best where the bulbs are hot and dry in summer, although it will tolerate low temperatures in winter.*

Wilson leapt clear—only for a falling rock to smash his leg. His leg was splinted with his camera tripod, and barely avoided complete amputation thanks to the skills of a medical missionary. Wilson was left with what he called his "lily limp".

Curving turk's-caps

The turk's-cap lilies have pendulous flowers with petals reflexed, or turned back, to form a shape resembling the turbans seen in Ottoman courts during the Middle Ages. In Europe, the common turk's-cap is *L. martagon*, which grows in mountain woodlands from western Europe across to central Siberia. It is typically a dull pinkish-purple, although white or wine-red flowers also occur. Despite their elegant appearance, they have an unpleasant scent and are best planted in the wild garden. Several other European turk's-caps have bright red flowers. They include the Greek *L. chalcedonicum*, introduced to western Europe in the 1500s.

In eastern North America, the common turk's-cap lily is the reddish-orange *L. superbum*. A magnificent marsh plant with each stem bearing up to 40 flowers, it can attain 10 feet (3m). *L. superbum* was sent by the pioneer American botanist and planthunter, John Bartram, to his correspondent John Collinson in London in 1735. We know that these plants did well, because three years later they were painted by botanical artist Georg Ehret.

The flowers of *L. superbum* and several other American turk's-cap lilies are abundantly spotted with maroon. In *L. pardalinum* from western North America, this feature has led to the name of "leopard lily." It is a clump-forming plant with stems not exceeding 6 feet (1.8 m), bearing relatively few flowers, but easily grown in damp, well-drained soil.

One of the earliest to be introduced to Europe—as far back as the 1600s—was the gold-orange *Lilium canadensis*, the "Canada lily." It was among the first plants that the French plantsman André le Nôtre selected for the gardens he designed for King Louis XIV. It thrives in lime-free soil enriched with leafmold, and in partial shade.

Tigers and devils

The "tiger lily," *L. lancifolium* (formerly *L. tigrinum*), is a species of turk's-cap with bright orange, purple-dotted flowers. It is an ancient garden plant in China, Korea and Japan,

REGAL ACCLAIM
The trumpet-shaped flowers of Lilium regale *(above) diffuse a strong scent, which contributed to its instant popularity when introduced from China in the early 1900s.*

MADONNA LILY
Lilium candidum *(left) was disseminated throughout the Mediterranean on Phoenician trading ships carrying purple dyes, Cypriot copper, and Egyptian linen from Asia Minor to the Strait of Gibralter and beyond.*

although it was cultivated not so much for its flowers, as for its edible bulbs—which are said to have a rather glutinous, starchy consistency when boiled. The tiger lily, known in Japan as *Oniyuri*, "devil lily, " was brought to western Europe in 1804, when bulbs reached Kew Gardens, near London. It was so easily propagated from tiny bulbs or bulbils in the angles of its leaf stems, that eight years later Kew distributed 10,000 bulbs.

The orange lily, *L. bulbiferum*, is a European species that can also be propagated easily in this way. Its orange, upward-facing flowers are the symbol of the Dutch Royal House of Orange, and the Orange Order of Northern Ireland, which commemorates the victory of William of Orange over deposed King James II in 1688. Orange is still the national color of the Netherlands, although when the Dutch revolted against their rulers in the 1780s, all orange flowers—including the lilies—were banished from gardens. (Oranges and carrots were also removed from vegetable stores and market stands.) The upright flowers of *L. bulbiferum* are now seldom seen in gardens. They have been superseded by hybrids such as *L. x hollandicum*, and cultivars like 'Enchantment', with up to 15 orange-red flowers on a more compact plant.

Intermediate in shape between the Turk's cap and trumpet lilies is the Himalayan *L. nepalense*, whose reflexed 6 inch (14cm) long petals are pale green with a large, central, maroon blotch. It is found in open places in the forests of Nepal and Bhutan and flowers during the monsoon rains. In cultivation it appreciates a lot of water during the growing season;

although the bulbs are hardy, the emerging shoots are vulnerable to spring frosts; in temperate gardens, this stunning lily is more safely grown inside or in a greenhouse.

TULIP MANIA

Ogier Ghiselin de Busbecq (1522-1592) was ambassador from Ferdinand I, the Holy Roman Emperor, to the Ottoman empire under Suleiman the Magnificent. While traveling to Constantinople (now Istanbul) in the early spring of 1554, he noticed some fine flowers in Turkish gardens. Among them were "those which the Turks call *tulipam* … admired for the beauty and variety of their colors." However, the Turkish word for a tulip is *lalé*. It seems that Busbecq misunderstood his interpreter describing the flowers as being like a *dulban*, meaning "turban"—and from a corruption of this come the name tulip and its botanical genus *Tulipa*. Surprisingly, this was the first time a knowledgable westerner had recorded the tulip, although several species grow wild in Greece and Italy.

Busbecq, a keen gardener, sent tulip bulbs and seeds to Vienna, where the plants were illustrated by the German naturalist Konrad Gesner in 1559. These first tulips in Europe were long-established, red-flowered cultivars from Turkey, rather than the true wild species, but they still bear their specific name in Gesner's honor, *T. gesneriana*.

The fame of the new plants spread rapidly throughout Europe. Within three years they were being grown in Antwerp, Belgium. The date of their appearance in the country now famed for their development, the Netherlands, is unknown, but was probably a few years after Belgium. By 1578, there were tulips in England.

Fanatical breeding

Busbecq may have introduced tulips to Europe, but the man who popularized them, in the late 1500s and early 1600s, was the Frenchman Carolus Clusius (1526-1609), who has been described as "the father of all beautiful gardens." Clusius was born in the town of Arras, but family conflicts caused him to travel widely. He acquired a medical training and also produced several botanical works, establishing his reputation as a serious gardener and botanist. In 1573, with the influence of Busbecq, he was appointed Prefect of the Imperial Gardens in Vienna, where he remained for 14 years. Clusius was particularly interested in

tulips. He began to grow them from seed—a delicate and dedicated process, since young plants may not flower for the first time until ten years of age.

In 1593 Clusius moved to Leiden, in the Netherlands. Here he established the *Hortus Academicus*, the first botanical garden to concentrate on ornamental plants, rather than medicinal ones. His private tulip collection also made the move to Leiden—and unwittingly founded the Dutch bulb industry.

Tulips were already growing in Dutch gardens at the time, but Clusius' plants were far superior. However, he was loath to part with any, so he asked extremely high prices for the bulbs. Local gardeners were either unable or unprepared to pay such a high premium, but they were determined to acquire some plants. So they broke into the botanic garden and stole most of Clusius' tulips. The progeny of his "liberated" collection started the Dutch tulip-growing industry—and also caused one of the strangest flurries of financial speculation that the world has seen.

The price of bulbs

Tulips have always been prone to infection by a virus that causes the flowers to "break" into streaks and speckles of different colors—especially if the background color is pale, when the marks may be red or pink. The virus can spread from plant to plant, and even in almost identical flowers of the same stock, it may cause a whole series of different-looking individuals. These "broken" tulips, although the result of disease, can be very attractive, and valuable.

In the early 1600s, Dutch gardeners began to spend ever-increasing sums on the finest broken forms. In 1623, a bulb of the cultivar 'Semper Augustus' sold for the then-fortune of thousands of guilders. This encouraged not only plant breeders, but also speculators, merchants, traders, financiers, and the nobility to try and earn some instant wealth from tulip dealing.

Soon bulbs were passing from one owner's hands to the next, often without being removed from their beds. Each buyer paid an ever higher price—and not necessarily in cash. The selling price for one bulb of 'Viceroy', white flowers streaked with bluish-pink, was "2 loads of wheat, 4 loads of rye, 8 fat pigs, 12 fat sheep, 2 hogsheads of wine, 4 barrels of beer, 2 barrels of butter, 1,000 pounds of cheese, a complete bed, a suit of clothes and a silver beaker."

The inflationary bubble burst in 1637;

A valuable export

Tulips have been portrayed in master works of art, traded as wealth, and finally become as closely identified with Holland as clogs and windmills. More prosaically, they are the foundation of a trade in bulbs worth $750 million to the Dutch economy each year. Some 44,000 acres (17,800ha) of the fertile, flat fields of Holland are devoted to the annual production of 9 billion bulbs, one third of which are tulips. Three billion tulips would be sufficient, if planted 4 inches (10cm) apart, to encircle the equator seven times! Lilies, gladiolus, and narcissus are next in order of importance, with smaller bulbs being grown in comparatively tiny numbers. Most Dutch bulbs are exported, with the United States, Germany, and Japan being the top importers. A large proportion are forced, effectively killing the bulb, to produce cut flowers, making the huge annual crop essential.

In spring, the bulb-growing areas are a major tourist attraction, enhanced by the bulb display garden at De Keukenhof near Lisse. Here 70 landscaped acres (28ha) are planted with six million bulbs, creating a dazzling display of color from March to May as "rivers" of blue grape hyacinths (*Muscari*) wind between banks of daffodils and tulips in all colors. Although Dutch growers produce bulbs of 3,500 different tulips, bright red- and yellow-flowered cultivars are the most popular, despite a general trend toward pastel shades in the garden. Response to fashions is not easy, however, as it takes between 12 and 25 years to produce a new cultivar in economic quantities.

Virus-induced beauty
The striping of the tulips in the picture (left) of the early 1800s, first published in Robert Thornton's 'Temple of Flora', was caused by a viral infection.

America's "first botanist"

John Bartram (1699-1777) was a Quaker farmer from Pennsylvania, who taught himself botany and adapted part of his farm to create America's first botanic garden in 1730. He also opened a nursery which supplied American plants and seeds to friends and customers in Europe, as well as to American president George Washington for his home, Mt. Vernon.

Between 1736 and 1766 Bartram went planthunting along the East Coast from Canada to Florida. The plants he found were kept in ox bladders filled with moss,

which kept them fresh. Bartram's Quaker connections put him in touch with Peter Collinson (1694-1768), who owned premises near the Pool of London— convenient for newly arrived consignments of plants. Although the two men never met, their relationship became one of the most fruitful partnerships in the history of botany. Through Collinson, Bartram introduced at least 200 species to Europe, including the American turk's cap lily, *Lilium superbum*, which flowered in Collinson's garden in 1738, and the Venus fly trap, *Dionaea muscipula*, as well as phloxes, asters, and golden-rods. Through Collinson's influence, in 1765 Bartram was appointed King's Botanist to King George III. The appointment seems to have been more of an honor than a financial benefit, since its salary of £50 (around 85 dollars) did not even cover freight costs.

The famous Swedish naturalist, Linnaeus, called Bartram "the greatest natural botanist in the world." Certainly his meticulous notes on the natural habitats of the plants he found made it easier for other gardeners to grow the plants he introduced.

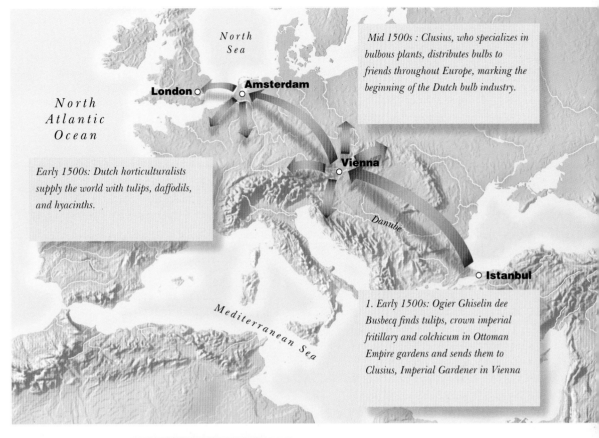

GOING DUTCH
The map shows the extraordinarily rapid distribution of tulips and other plants from the Ottoman Empire throughout Europe during the 16th century.

Mid 1500s : Clusius, who specializes in bulbous plants, distributes bulbs to friends throughout Europe, marking the beginning of the Dutch bulb industry.

Early 1500s: Dutch horticulturalists supply the world with tulips, daffodils, and hyacinths.

1. Early 1500s: Ogier Ghiselin dee Busbecq finds tulips, crown imperial fritillary and colchicum in Ottoman Empire gardens and sends them to Clusius, Imperial Gardener in Vienna

MASS PRODUCED
The long-established cultivar 'Keizerskroon' (left), still a staple of the Dutch bulb industry, was first recorded in 1750.

suddenly there were more sellers than buyers, the market collapsed, and the strange episode of "tulipomania" was over.

More modern tulip cultivars

Most tulip cultivars currently available commercially were selected during the present century. An exception is the fine red and yellow 'Keizerskroon', meaning "Emperor's crown," which dates from around 1750. One of the most commonly seen today is 'Apeldoorn,' with bright red flowers. It was introduced only in 1951, and is the product of a cross between old garden tulips and *Tulipa fosteriana*, a vivid scarlet species from Central Asia. Other forms, such as 'Queen of Night' and 'La Tulipe Noire' are so deep chocolate-purple that they appear almost black. The similarly dark 'Black Parrot' also has twisted and fringed petals—caused, like the streaking of old cultivars, by viral infection.

The wild species of *Tulipa* are less common in gardens today. Most came from Turkey and Central Asia. *T. clusiana*, named after Clusius himself, was sent to him at Leiden in 1607. It has elegant, slim flowers less than 2 inches (5cm) long, cherry-red on the outside and opening to reveal a white interior with a handsome maroon

STATELY ELEGANCE
Fritillaria imperialis, the crown imperial, was introduced from Turkey to Europe in 1576 by Carolus Clusius.

blotch at the base. It never sets seed, but spreads by stolons (long side branches or shoots, similar to runners), and enjoys a warm, well-drained position in full sun, as do all tulips. Indeed, most tulips are best lifted and replanted every year, to give them a dry summer rest. But the most vigorous, such as *T. fosteriana* 'Purissima,' a reliable clone with huge, creamy-white flowers, can be left in the ground.

FRITILLARY: CROWNING GLORY

The partnership of Busbecq and Clusius was responsible for importing and developing not only tulips, but many other species of bulbous plants from Turkey, which greatly enriched the gardens of western Europe. After a period of relative stagnation during the Middle Ages, they provided gardeners with a treasure-trove of plants for breeding and selection.

One of Clusius' best-known introductions, and still one of the most statuesque of hardy plants, was the crown imperial, "corona imperialis." This is the handsome, red- or yellow-flowered *Fritillaria imperialis*, introduced from Turkey to Vienna in 1576. John Parkinson, the English gardener, wrote in 1629 that it was the finest of lilies, and Shakespeare

with most members of its genus, the crocus-like flowers appear before the leaves in late summer—a habit which has given them the common name of "naked ladies." *C. speciosum* from the Caucasus Mountains of western Asia is a larger, darker-flowered species.

HEMEROCALLIS: THE DAY LILY

The genus *Lilium* contains about 100 species of true lilies. But the common name "lily" has been applied to various other plants, especially those with trumpet-shaped flowers. Some are members of the Lily family, such as *Gloriosa superba*, the gloriosa lily, whose red and yellow flowers are the national floral emblem of Zimbabwe. Other species are not within the Liliaceae. The arum lily, *Zantedeschia aethiopica*, is a member of the Araceae.

"Day lilies" are members of Liliaceae, but of the genus *Hemerocallis*. These most tolerant of plants, which thrive from the Arctic to the tropics, are named from their short flowering period. Breeders have been unable to persuade the flowers to remain open for more than one day, the habit which gives rise to the scientific name, *hemero-* meaning "day," and *-callis*, "lily".

Hemerocallis fulva, the common tawny-orange species, was carried westward from its native China and Japan by early traders (see map on p. 111). In China it is called *hsuan t'sao*, "flower of forgiveness," because it supposedly cures sorrow and grief by inducing memory loss; more prosaically, its petals are commonly eaten in soups and stews.

The day lily's westward travels continued with settlers, and eventually became naturalized. Today, breeders in the U.S. have actively raised many new cultivars in a wide range of colors—from pure white, through orange and pink shades, to dark mahogany-brown. The narrow petals of the wild species have also been broadened and overlapped.

The American Hemerocallis Society is one of the largest of the country's specialist flower societies, with more than 10,000 members. The Hemerocallis Society is internationally responsible for registering new day lily cultivars, amounting to hundreds each year, adding to the 40,000 or so already recorded. One of the foremost day lily breeders is Pat Stamile, who gave up his teaching post in New York to grow the plants full time in Florida. His pink 1989 cultivar 'Strawberry Candy' and 1987 'Wedding Band', near white with a gold edge, are among the most popular day lilies in the entire U.S.

mentioned it in *A Winter's Tale*. The pendulous flowers are held above the main stem leaves, and are themselves surmounted by a "crown"—a tuft of leafy bracts (leaf-like parts). However the name crown imperial comes not from this structure, but from the plant's early cultivation in the imperial gardens of Vienna.

Fritillaria imperialis is the largest of the many species in the genus, named from the similarity of their angular bell-like flowers to a *fritillus*, the pot-like shaker from which the Romans threw dice. The snakeshead fritillary, *F. meleagris,* is a northern European species of damp meadows, sometimes growing in enormous numbers. This common name derives from its unopened buds, which look distinctly scaly and reptilian. In cultivation, the snakeshead fritillary can be grown in grass, on the rock garden or in light shade. Its purple flowers are regularly checkered with paler marks, leading to its species name *meleagris*—meaning a guinea-fowl, in allusion to the bird's similarly checkered plumage. White forms, in which the checkering may be outlined in green, are quite common in wild populations.

Another plant introduced by Clusius, and still propagated, is *Colchicum byzantinum*, a pale pinkish-mauve meadow saffron from Turkey. As

CHECKERED SHADES
Although a protected species in Britain, Fritillaria meleagris *(above) does naturalize quite easily, occasional white forms appearing among the more usual shades of purple.*

TRANSFORMATION
During the 20th century, breeders have improved the wild Hemerocallis fulva, *or "ditch lilies" (right) from eastern Asia and transformed their colors from the wild plants' yellows and oranges.*

ANCIENT PLANTS OF PARADISE
GARDENS AND MEDITERRANEAN
EVENINGS, DELICATE OF FORM AND EXQUISITE OF SCENT.

OLIVES, PLUMBAGOS, AND CONVOLVULUS
(OLEACEAE, PLUMBAGINACEAE, CONVOLVULACEAE)

Plumbago auriculata

The olive family, Oleaceae, contains trees, flowers, and shrubs which have been appreciated and cultivated for thousands of years, including the olive itself, sweet-scented lilacs (*Syringa*), and jasmines (*Jasminum*). Many hail from the Middle East and Asia, but are now grown in many parts of the world, especially in warmer climates. While the olive has been cultivated as a food crop in the Mediterranean countries for over 5,000 years, lilac and jasmine

Key

Extent of the Islamic world

Ottoman Turks specialize in bulbous plants which become a major source for Western gardens during the 1500s.

The Moors imprint Islamic style on the gardens of North Africa and Spain, from the 7th to 15th centuries.

LILAC TIME
Syringa x *persica* (left) was the first lilac to reach Europe from Asia. Its Persian name *nilak* was corrupted into "lilac," and then gave its name to the color.

Arctic Circle

North Atlantic Ocean

Danube

Bl

○ Toledo
○ Cordoba
Seville ○ ○ Granada

Mediterranean Sea

Damas

Jerusal

○ Fez

Tropic of Cancer

Niger

Nile

Origins of Islam 7th C A.D. The Prophet Muhammed promises Paradise, portrayed as a cool garden, to the faithful.

○ **Karbala**

Equator

Congo

are traditional plants of the pleasure gardens of Persia (Iran), dating back to 2,000 B.C.

LILACS IN PARADISE

The scientific name *Syringa* is derived from the Greek word *syrinx*, meaning "pipe," because lilac stems are hollow and pipe-like, although filled with spongy pith.

In Greek legend, Syrinx was a maiden who escaped the unwanted carnal attentions of the god Pan, when she was suddenly turned into a reed. Pan then crafted his well-known musical pipes from the reed, and these are still technically known as a syrinx.

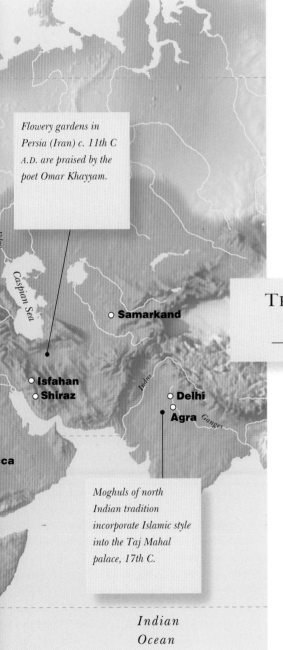

Flowery gardens in Persia (Iran) c. 11th C A.D. are praised by the poet Omar Khayyam.

Moghuls of north Indian tradition incorporate Islamic style into the Taj Mahal palace, 17th C.

Caspian Sea

Volga

Samarkand

Isfahan

Shiraz

Indus

Delhi

Agra

Ganges

cca

Indian Ocean

HEAVEN SCENT

Jasminum *officinale* *(above) was often grown in Islamic gardens, where the perfume from the white flowers contributed to the heavenly atmosphere.*

THE WORLD OF ISLAM
—gardens in paradise

Gardens in the Islamic world symbolized heaven and were called pairidaeza, which literally means "enclosed park," and from which the word paradise is derived. The garden culture—and the plants which were cultivated—spread as the Islamic world expanded from the 7th to 15th centuries, taking species such as lilac and jasmine westward.

The Persian lilac, *S.* x *persica*, is a hybrid of uncertain ancestry. The Moors took it from its western Asian home as they conquered North Africa and Spain from the 8th century. It was then brought to northern Europe by the English plantsman John Tradescant the Elder in 1621. Tradescant was always ready to seize the chance of exploring new areas. He had joined an expedition sent against the so-called Barbary pirates, who were based in Algiers and terrorized Mediterranean shipping, by suggesting that he might be a useful gunner. His claim was treated with the scepticism it probably deserved, but he did gain the opportunity to visit North Africa, and brought back the lilac.

Syringa x *persica* has tubular flowers which are, indeed, lilac in color. It forms a small, dense shrub, but is now less commonly seen than the common lilac, *S. vulgaris*, and its hybrids, from southeastern Europe. The name Persian lilac is also given to *Melia azedarach*, known as the "bead-tree" or "chinaberry," from the family Meliaceae, which also has bluish, scented, *Syringa*-like flowers. The German word for this tree, often found on roadsides in the Mediterranean countries, is *Paternosterbaum* because of the use of its five-sided seeds as rosary beads.

Syringa vulgaris, which can grow into a small tree, bears larger heads of flowers than the *S.* x *persica*. Also the flowers, although typically lilac, can vary in color from white to deep red-purple, and even pale yellow in the cultivar 'Primrose'. Like all lilacs, they are noted for their delightful

perfume. Many cultivars of *S. vulgaris* were raised by the Lemoine nursery of Nancy, France, during the late 1800s and early 1900s. The nursery's achievement is commemorated in the widely grown double-white cultivar 'Mme. Lemoine' of 1894. Other lilac species, such as the Chinese *S. oblata* and *S. reflexa*, have also been used to create hybrid lilacs with different flower shapes and colors. However, they have not achieved quite the same degree of popularity as *S. vulgaris* cultivars.

Jasmines galore

John Tradescant's Barbary expedition also resulted in the arrival of the first plants of the vigorous and entwining summer-flowering jasmine, *Jasminum officinale*, in England. The sweetly scented, white flowers of this climber had been appreciated for centuries in gardens throughout the Islamic world. In fact traders had originally obtained the plants from their native western Himalayas. The English name derives from the Persian one, *yasmin*.

Jasminum officinale is the only white-flowered, scented species which is hardy. It can survive where the temperature falls to 0°F (minus 18°C). The even more floriferous and strongly-scented *J. polyanthum*, from southern China, must be protected from frost—although it flowers only after a sudden drop in temperature. Its five-lobed flowers are white with a pink reverse, and in subtropical or mediterranean climates they usually appear in late winter or early spring. This jasmine can be grown in more northerly areas in a glasshouse or conservatory, but in confined spaces the scent may be cloying—it is best experienced wafting on the warm air of a tropical evening.

Winter exuberance

The winter-flowering jasmine, *J. nudiflorum*, bears yellow flowers on leafless branches, as the specific Latin name suggests. The flowers are one of the great joys of a temperate winter, and while they are killed by frost, they are soon replaced by others. Although the plant is at its most prolific and eyecatching in a sunny position, it will also perform reasonably well in a northfacing position.

The species was discovered near Beijing, China, by botanist Alexander von Bunge (1803-1890), as he accompanied a Russian ecclesiastical mission to the Chinese capital in 1829-30. It was introduced to Europe in the 1840s by Robert Fortune, the Royal

WINTER SUN
Appearing on bare twigs,
the yellow flowers of
Jasminum nudiflorum
(above) brighten the dark
days of temperate winters,
with new blooms replacing
those lost by frost.

ANCIENT HERITAGE
Ubiquitous in the
Mediterranean and
worldwide symbol of
peace, the silvery leaved
olive tree (Olea europaea)
plays an incidental
ornamental role as an
attractive evergreen in
warm climates (right).

Horticultural Society's collector in China. The flowers that impart the fragrance to jasmine tea come from another Chinese species, *J. paniculatum*, but this is seldom seen as a cultivated plant.

Olive: symbol of goodwill

The olive itself, *Olea europaea*, has been cultivated in the Mediterranean area for almost the entire time span of recorded history. Its role as a symbol of peace and goodwill, was especially significant in Ancient Rome, when adversaries would "offer the olive branch" to begin negotiations. Today it is chiefly grown for the oil expressed from its fruits. The gnarled appearance of the oldest trees, such as those in Jerusalem's historic Garden of Gethsemane, may not be achieved in the average gardener's lifespan! But as a drought-resistant tree it is well suited to any type of sunny, warm, and dry climate, from California to southeast Australia, especially where there may be restrictions on water use. In frost-prone areas, young specimens can be grown as potted ornamentals.

Olive flowers are not as significant as the black or greenish oval fruits. Freshly picked olives need prolonged and careful treatment before they are edible.

DAWN-SKY BLUE
A native of South Africa, the shrubby <u>Plumbago auriculata</u> is grown in frost-free areas or conservatories for glorious abundance of soft blue flowers.

PLUMBAGOS: THE LEADWORTS

The leadworts, with their heavenly blue flowers, are members of family Plumbaginaceae. The flowers are five-lobed like those of the jasmine, but they lack fragrance and climbing ability. The family and generic names come from *plumbum*, Latin for "lead," since the European species, *Plumbago europaea*, was reputed to cure lead poisoning. Plumbagos are found in warm areas around the world, the most commonly grown species being the South African *P. auriculata*, or Cape leadwort. Its freely produced flowers make it a decorative shrub in mediterranean-type climates, or in the conservatories of cooler

localities. In more temperate regions, where *P. auriculata* cannot survive outside, substitutes include its near relatives, *Ceratostigma plumbaginoides* and *C. willmottianum*. These low shrubs, originally from China, bear bright blue flowers in late summer, often continuing into fall, making a striking color combination as the leaves turn red.

Ceratostigma willmottianum was collected in 1906 by Ernest Wilson, and named for one of his expedition sponsors, the redoubtable Miss Ellen Willmott (1858-1934) of Warley Place, Essex, England. Miss Willmott gardened in the grand manner, at one time employing more

*P*ARADISE, AS ENVISIONED IN *THE KORAN*, IS A COOL, SHADY GARDEN FLOWING WITH STREAMS AND RIVERS—AN ENTICING CONTRAST TO THE DESERTS WHICH GAVE BIRTH TO THE ISLAMIC CULTURE.

The Court of the Lions

GARDENS OF ISLAM

In the Islamic world of the Middle Ages, stretching from Moghul India to southern Spain, the wealthy strove to create an earthly version of paradise—a vision still to be seen in Granada, Spain in the gardens of the Alhambra and Generalife, which have survived five centuries of Christian rule. The Alhambra is a fortress planned for elegant living. It occupies a hillside above Granada with spectacular views to the valley beneath and the snow-capped Sierra Nevada beyond. Conversion of the old fort to palace and gardens was begun in the mid-13th century A.D. by Mohamed ben Al-Ahmar, whose Nazarite dynasty occupied the Alhambra and adjacent Generalife palace for 250 years.

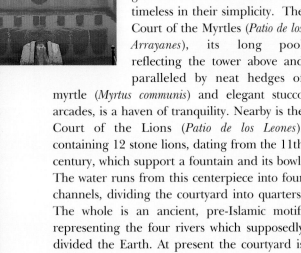

Within the Alhambra the gardens are restrained and timeless in their simplicity. The Court of the Myrtles (*Patio de los Arrayanes*), its long pool reflecting the tower above and paralleled by neat hedges of myrtle (*Myrtus communis*) and elegant stucco arcades, is a haven of tranquility. Nearby is the Court of the Lions (*Patio de los Leones*), containing 12 stone lions, dating from the 11th century, which support a fountain and its bowl. The water runs from this centerpiece into four channels, dividing the courtyard into quarters. The whole is an ancient, pre-Islamic motif, representing the four rivers which supposedly divided the Earth. At present the courtyard is covered with gravel and has a few formal clipped shrubs, but in Moorish times it was probably fully planted.

The gardens around the Alhambra and at the Generalife have a more exuberant formality appropriate to the wider area available and more distant relationship to the buildings. Within the Generalife, formality reasserts itself in the Court of the Long Pond (*Patio de la Acequia*). Here, a canal containing many fountains runs between richly planted beds in a narrow courtyard. This is intended to mirror the Koran's description of Paradise as a "dark green and luxuriant garden with a river running through it, planted with fruit trees, pomegranates and palms."

Aromatic introductions

The Moors brought many plants to Spain, including oranges and lemons, apricots, almonds, and many herbs, as well as adopting wild plants for their gardens. Strongly scented plants were favored, adding perfume to the cool air of paradise. The modern planting at the Alhambra and Generalife is underpinned by the traditional cypresses, bays, and other evergreens of the Mediterranean basin. In complete contrast, the many watercourses and fountains are overhung by the dainty maidenhair fern (*Adiantum capillus-veneris*). The soothing water and scented shade are a world away from the surrounding dry, Mediterranean landscape.

COOL SYMMETRY
The map, the Court of Myrtles (left), and the garden (right) display the Islamic sense of order and geometry, the important, and the distinctly Spanish characteristic of several, linked enclosures.

1. *Court of the Myrtles*
2. *Court of the Lions*
3. *Gardens of the Partal*
4. *Palace of Charles V*

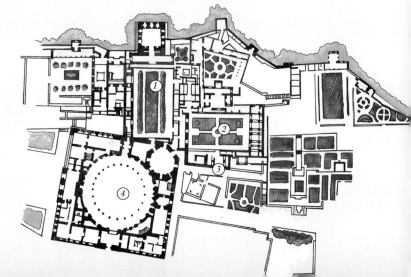

than 100 staff to maintain her plant collection. But this outlay sapped her fortune and she died in poverty. However she is also remembered from her habit of scattering the seeds of *Eryngium giganteum*—a thistle-like umbellifer plant with silvery-white bracts—on other people's property. These grew a year or two later, and the species became known as "Miss Willmott's ghost."

Dried flowers of the coastline

Many members of the Plumbaginaceae are found in coastal habitats around the world. To cope with salty spray, they have evolved specialized glands to excrete excess salt from their tissues. One is the small, tufted *Armeria maritima*, known in England as thrift—although it seems to have no connection with careful financing! Thrift is grown as a rock garden or edging plant, its tight cushions of small narrow leaves producing flowering stems topped with heads of pink or red flowers.

Armeria maritima was known to the Ancient Greeks as *statice*, a reference to its potential for binding, or making static, shifting sand dunes. But the common name of "statice" has since been transferred to the similarly binding sea lavenders of the genus *Limonium*. In both genera, the calyx becomes dry after flowering, and serves as a parachute for dispersing the seeds. But in most *Limonium* species, in contrast to *Armeria*, the calyx is brightly colored, and remains so when dried, making it a popular dried flower—although the sepals of the calyx are often mistaken for petals. The species

EXTENDED LIFE
*Swathes of Limonium
sinuatum (right) color
meadows along the shores
of the Mediterranean in
summer, and retain their
colors indefinitely as
"everlasting" dried
flowers if kept out of
direct sunlight.*

usually grown, as an annual, for drying in this way is *L. sinuatum*, a perennial from Mediterranean coasts. The wild plant has a white or blue calyx, but selection has widened the color range to include pink and purple. Yellow and orange shades are derived from the closely related *L. bonduellii* which comes from North Africa.

CONVOLVULUS: GLORIOUS WEED

The bindweed family, Convolvulaceae, includes some of most rampant weeds facing the gardener: *Convolvulus* and *Calystegia*. But the genus *Ipomoea*, a tender, tropical plant with blue, trumpet-shaped flowers called "morning glories" is much appreciated. The common name refers to the ephemeral flowering habit; the blooms open soon after sunrise, but fade by the afternoon.

In warm temperate gardens, the most familiar morning glory is the annual *Ipomea tricolor*, whose flowers may be white, red, or pale blue. This native of Central America is now widely naturalized in the tropics. A similar species, also annual but hardier, is *I. nil*, with slightly smaller flowers varying from red and blue, to chocolate brown and deep, velvety purple-black. It may be sown outside in cooler temperate gardens in spring.

Ipomea indica, sometimes known as *I. learii*, was a perennial morning glory favored in conservatories and greenhouses in the late 1800s. It is now seldom seen in temperate regions, but it has been widely planted in the tropics. In India it was often established by railroad stations, and is known as the "railway creeper." *I. indica* can be rather rampant, scrambling through low trees to create curtains of growth, covered each morning with deep blue flowers which fade to magenta by midday.

An exception to the ephemeral flowering habit is *I. alba*. Its white flowers, up to 6 inches (15cm) wide, open at dusk. This has earned it common names such as "moonflower" and "*belle de nuit*" (night beauty). It is found both wild and cultivated throughout the tropics. The genus also includes an important food crop, *I. batatas*, the sweet potato.

Seeds of many morning glories, such as *Ipomoea tricolor*, contain derivatives of the chemical lysergic acid, also known as LSD. Their hallucinogenic and narcotic properties were known to, and used by, the Aztecs. For this reason, cultivation is banned in some regions, such as parts of the U.S.

$\mathscr{F}$AMILIES FAMILIAR IN THE WILD
FROM THE AMERICAS TO THE
ANTIPODES, WHICH HAVE BROUGHT ELEGANCE OF FORM, VIBRANT COLORS,
AND OFTEN LONG FLOWERING SEASONS TO CULTIVATION.

$\mathscr{F}$UCHSIAS AND $\mathscr{O}$LEANDERS
(ONAGRACEAE, APOCYNACEAE)

Plumeria rubra 'Frangipani'

The genus *Fuchsia* contains mostly trees and shrubs native to South America, which have since been bred for their distinctive blooms and long flowering season. The scientific name commemorates the successful German physician and botanist, Leonhard Fuchs (1501-1566). Yet Fuchs himself never saw one of these plants, having died more than a century before living specimens arrived in Europe.

The naming came about because Fuchs combined his two main interests to produce one of the best known and most important herbal publications of the 1500s, *De Historia Stirpium* (1542) which coupled accurate text with fine illustrations. Its contribution to botanical knowledge was recognized in the next century by French planthunter Charles Plumier (1646-1706) who honored many famous European botanists, including Fuchs, with his discoveries in tropical America. Plumier studied, described, and named the fuchsia he discovered—possibly the plant we now call *F. coccinea*—in 1703. But, like Fuchs, he too never saw living specimens arrive in his native Europe.

Two species of fuchsia did finally arrive in the Royal Gardens at Kew, near London, in 1788, delivered by a Captain Firth, who transported them from South America. But by the time a London nurseryman, James Lee, began to distribute the plants to a wider public, the story of their introduction had altered. Lee claimed that an ordinary sailor had brought back a fuchsia plant as a souvenir from South America for his aged mother, who loved flowers. The old lady grew this treasure in the window of her home in Wapping, in East London's docklands. Lee said that he had heard rumors about the new flower, so he went to Wapping and negotiated with the old lady, paying her 80

ONAGRACEAE IN THE AMERICAS

Clarkia amoena *grows, like many annual species of California and other western States, as annuals, germinating and growing during the wet winters, then* *flowering in spring before the soil dries out. In a good year with a mild, wet winter, they and other flowers like the California poppy (Eschscholzia) cover the hills in a mosaic of color. Other Clarkia species have narrower, more starry flowers, but all bloom in shades of yellow, white, pink, and purple.*

Pacific Ocean

Zauschneria californica *has long, funnel-shaped flowers adapted for pollination by hummingbirds, which are attracted to the color red. Found on dry hillsides in the western States and northern Mexico, it blooms in late summer when few other* *plants are in flower. In cultivation, it will continue to bloom until the first frosts. Where winters are severe it is best over-wintered under cover. Zauschneria is included in the genus Epilobium by some botanists.*

guineas for the mature plant and two young specimens. Lee raised an enormous number of plants from the easily-rooted cuttings and sold them to a fuchsia-hungry public at great profit. But was his tale true? Or was it the embellishment of a salesman to conceal a less-than-scrupulous acquisition from Kew?

The hardiest fuchsia

One of Firth's original two fuchsias was *Fuchsia magellanica*, a native of the cool, moist forests of southern Chile, which extend as far as Tierra del

Fuego at the extreme southern tip of South America. In the wild it grows as a woody shrub, reaching 16 feet (5m) or more. In cultivation it tends to form a small dense shrub, which can be used for hedging, and it is planted for this reason in Ireland and western Scotland. Its original southerly distribution makes it the hardiest fuchsia, surviving temperatures as low as 14°F (minus 10°C). Where winters are this severe, it grows as a woody-based herbaceous plant, producing new shoots from the rootstock each spring.

The flowers resemble elegant ballerinas—a red calyx "body" spreading above the purple petals of the flowing "skirt," with the anthers and style emerging like slim legs below. After pollination, they develop into tasty black berries that make excellent preserves. Many variants are known, with blooms of different sizes and shapes. The cultivar 'Riccartonii' has rounded flowers. Var. *molinae* (formerly *alba*) has smaller blooms of palest shell-pink. It comes from the island of Chiloe, off the Chilean coast, which is noted for its exceptionally wet climate.

The complexity of fuchsias

There are more than 100 species of fuchsia, and many followed the original two to Europe, chiefly during the early 1800s. They were immediately hybridized with each other, and the development of the various breeding lines became so complex and confused that it is impossible to document or unravel. Probably, at least ten species have contributed to the vast range of more than 8,000 modern hybrids and cultivars. *F. fulgens*, introduced from Guatemala during the 1840s by Theodor Hartweg (1812-1871), was a particularly significant species. Hartweg traveled widely in western America,

Oenothera biennis, *the evening primrose, opens its large yellow flowers as the light fades, attracting night-flying moths by their color and perfume. Early next morning the flower shrivels, to be replaced by others opening from the numerous buds. Other species have white flowers, while a few are pink. All are native to the Americas, but evening primroses are now found in many parts of the world where they have escaped from gardens.*

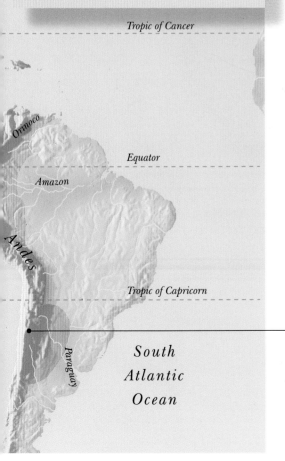

Tropic of Cancer

Orinoco

Equator

Amazon

Andes

Tropic of Capricorn

Paraguay)

South Atlantic Ocean

Key

Distribution of *Oenothera*

Distribution of *Clarkia*

Distribution of *Fuschia*

Fuchsia magellanica *is an elegant, small-flowered wild species from the wet forests of the west coast of South America. One of the first species to be introduced to gardens, and the hardiest of all, it was an important parent of many of the large-flowered hybrid fuchsias. Many cultivars, including several with variegated or golden leaves, have been selected. In Scotland and Ireland, where it flourishes in the humid climate, it is often planted as a wind-resistant hedge.*

UNUSUAL POLLEN
The prostrate stems of
Fuchsia procumbens
(above) bear small,
upward-facing flowers
which are followed by
large red berries. Blue
pollen is an unusual
feature of this plant.

COLOR MATCH
A large-flowered Fuchsia
hybrid (left) shows its
descent from F.
magellanica in its purple
and red coloration. Many
large-flowered fuchsias
are tender, and need to be
kept frost-free in winter.

collecting plants for the Horticultural Society of London. He was responsible for the introduction of many tender plants to Europe, including various orchids and cacti, as well as fuchsias.

Fuchsia fulgens has elongated, narrow petals of bright vermilion, with paler, green-tipped sepals. It has passed these characters to its many descendants, including the widely grown 'Thalia', with its spectacular combination of bronze foliage and orange flowers. However, *F. fulgens* and its descendants must be over-wintered in frost-free conditions.

Antipodean specialities
Most fuchsias originate in South America. Their mainly bright red flowers probably evolved to attract hummingbirds, as pollinators. In New Zealand, hummingbirds are lacking, and the few species of fuchsias there tend to have flowers in curious combinations of browns and greens. An example is the largest member of the genus, *F. excorticata*, reaching 40 feet (12m), with shaggy bark peeling from the trunk. Its somber flowers are enlivened by blue pollen, which was used by Maori ladies as face powder. They called the plant *kotutukutu*.

The smallest fuchsia species, *F. procumbens*, has similar flowers, but these are produced from trailing, wiry stems that hug the ground. *F.*

procumbens plants are now rare in the wild, confined to a few rocky outcrops along New Zealand shores. However it is common in cultivation, especially as ground cover. Both these largest and smallest species withstand a few degrees of frost, but do best in milder climates, especially in coastal districts.

EVENING PRIMROSES
The fuchsia family, Onagraceae, has representatives throughout the world, although like the fuchsias themselves, the majority are found in the Americas. One of these all-American genera is *Oenothera*, the evening primroses. The most familiar species, now worldwide, is *O. biennis*. Named after its biennial growth, it reaches five feet (1.5m) in height and bears a constant succession of pale yellow, fragrant flowers throughout the summer. The flowers open in the early evening, as the common name implies, releasing their pleasant fragrance to attract pollinating moths.

In 1619, this evening primrose was brought from Virginia to Italy. It was known in England by 1621. Today it behaves as a wild pioneer in many parts of Europe, favoring freshly disturbed ground. Its seeds, like those of poppies, can survive for decades. Experiments show that they retain their viability even when buried in soil for more than 80 years, waiting for the exposure to light caused by a disturbance such as digging.

In recent years, *O. biennis* has become a commercial crop. Some authorities recognize that the oil extracted from its seeds has beneficial medicinal properties, especially for female menstrual problems.

California specialities and "fireweed"
The fuchsia family includes the clarkias, a genus of wild flowers which are natives of western North America. In California, *Clarkia unguiculata* is known as the "mountain garland" and features prominently in spring displays of local wild flowers. The genus name commemorates the explorer Captain William Clark (see panel, page 126). The common name of clarkia is given mainly to cultivars of *C. unguiculata*, which has small, rounded flowers held in elongated spikes. In gardens, the cultivated forms are usually double or have expanded, frilled petals, and often bring their shades of pink to summer bedding displays.

The related *C. amoena* is known by its common name of "godetia," which was derived

Presidential commission

The explorers Captain William Clark (1770-1838) and Meriwether Lewis (1774-1809), were the first to cross the North American continent from east to west, in 1804-06. Their expedition was commissioned by President Thomas Jefferson, with the principal aim of finding a route across the continent by river, lake, and other waterways. He also charged the two men with making a full contribution to the knowledge of North American natural history. To this end, they made a collection of minerals, animals, and plants—although neither was a botanist.

As a result of their endeavors, the genera *Clarkia* and *Lewisia* (in the family Portulacaeae) were named for them. Lewisias are fleshy perennials from the mountains of western North America. Many have handsome, glistening flowers with two-tone petals, and all need special care mainly suiting alpine enthusiasts. The species from the original expedition was *Lewisia rediviva*, from the Rocky Mountains. The name comes from its ability to revive after long periods of drought—even after boiling! It is known locally as bitterroot, and despite its sharp taste, it has been eaten by Native Americans of the region.

The common name has also lent itself to the Bitterroot Range, which divides Idaho from the state of Montana—where this historic species is the State flower.

Lewisia rediviva

MEDITERRANEAN SHOW
The showy flowers (right) of oleander (Nerium oleander) are a characteristic feature of mediterranean climates. The flowers are followed by long, narrow pods full of silvery-plumed seeds, which drift away in a breeze.

from its former genus name, *Godetia*, before its transfer to *Clarkia*. This has much larger, upright-facing flowers on a compact plant. In Germany it is called *sommerazalee*, meaning "summer azalea," because its flowers are often bicolored with a darker spot in the center of each petal, resembling true azaleas. Many cultivars of this species have been developed, with flowers ranging from white through apricot to dark red. Like most clarkias, they grow best in full sun and well-drained soil—their natural habitat in the dry Californian hills.

Another plant from the hot, dry hills of California is *Zauschneria californica*, often called the "California fuchsia." Its brilliant, vermilion flowers, borne on a low, shrubby bush, are a favorite with humming birds. *Zauschneria* is a variable plant in the wild and several cultivars have been selected. One particularly hardy cultivar bred in the National Botanical Garden at Glasnevin in Dublin, Ireland, is called 'Dublin' and produces abundant flowers in late summer and early fall when bright color in the garden is very welcome.

Some botanists include *Zauschneria* in the genus *Epilobium*—the willowherbs—to which it is undoubtedly closely related. Many willowherbs are irritating weeds with dingy flowers, but a few, though potentially weedy, are decorative. Among these is *E. angustifolium*, the rosebay willowherb, which is found wild throughout the northern hemisphere. The North American name of "fireweed" comes from its habit of appearing in large quantities a year or two after forest fire. It became famous in Britain during the Second World War when it covered areas of London devastated by the Blitz with its bright, magenta-pink flowers. A white-flowered form is easier to place in cultivation, but is equally invasive. In Siberia, where the fermented leaves are used as a tea substitute, it is called *Ivanchai*—"John's tea."

OLEANDERS AND RELATIVES

The oleander, *Nerium oleander*, shares the common name of "rose bay." It is in the family Apocynaceae, all members of which contain a white latex or sap, which flows freely when a stem or leaf is damaged. The sap can be highly poisonous: chewing a single leaf of oleander can kill within 24 hours. However, it is an attractive shrub for warmer climates. In the wild, oleanders grow in the river beds of the Mediterranean area, where they bear single, five-lobed flowers of pale pink. Breeding has

created cultivars in shades of pink, red, white, even cream, and many with double flowers.

Oleander performs best where water is abundant. But it has evolved great tolerance, since its natural riverside homes often dry out in summer. So it can cope with dryness and also neglect—it even flowers freely on the central reservations of highways in warmer regions, such as California and South Africa. It can survive a few degrees of frost, too. But in northern North America and northern Europe, oleander is best grown as a pot plant for summer display.

The perfume of *Plumeria*

Frangipani was an Italian perfumier who lived during the 1100s A.D. He is credited with creating a heavenly scent by combining several essential oils. This fragrance was favored by the immensely influential Catherine de Medici, and eventually took on its creator's name. When early European settlers reached the Caribbean,

they smelled an almost identical scent emanating from the delicate, five-lobed flowers of a squat, thick-branched tree. The trees, which comprise a small and mainly tropical genus, *Plumeria*, in the family Apocynaceae, became known as "frangipanis." The genus is named after French planthunter Charles Plumier, and is one of the many genera he discovered while exploring the flora of the Caribbean and Central America in the late 1600s, on behalf of King Louis XIV of France. Two species, *P. rubra* with red, pink, or white flowers, and *P. obtusa* with white, yellow-centered flowers, are grown in tropical gardens.

Frangipani trees have thick, waxy-barked branches that shed their leaves and go dormant in the dry season. Yet a branch that is cut from the tree can later "resurrect," sprout new growth and flower. This feature has permitted the spread of frangipanis throughout the tropics, as dry cuttings. The apparent "resurrection" of dry cut branches has led to frequent planting of frangipanis in Buddhist and Muslim burial places, and near temples in India, they provide a ready supply of offerings. Frangipanis are also known as "pagoda trees." Interestingly, Plumier himself was a member of an unusual religious group, the Order of Minims. Members were required to eat a strict vegan diet and give their lives solely to the service of God. He might be interested to know that his namesake plant is important as a symbol to various religions.

Strong stems: periwinkles

Most members of the Apocynaceae are tropical, but a few occur in temperate areas. These include the periwinkles, *Vinca*, whose tolerance of dry shade makes them suitable as ground cover beneath trees. Species commonly planted for this purpose are *V. minor* and *V. major*, which are known, as one might suspect, as the lesser and greater periwinkles. The lesser *V. minor* is more restrained and typically bears many small pale-blue flowers in early spring. Cultivars sport white and red-purple flowers and even variegated leaves.

"Periwinkle" is a corruption of the Latin *pervincula*, "around band," a reference to the plant's strong, wiry stems. In Europe, the lesser periwinkle has rather morbid associations. It was fashioned into wreaths for the head of condemned criminals on route to execution, and in Italy it was known as *fiore di morte*, "flower of death," from its use as decoration on the funeral biers of deceased children.

One of the most admired cultivars of the lesser periwinkle, with rounded, large flowers, is 'La Grave'. This is named, not from its funereal associations, but after the French town where it originated. However, the plant which was developed as the cultivar was discovered in the town's cemetery.

The anti-cancer periwinkle

The Madagascar periwinkle was once included in the genus *Vinca*. It is now known as *Catharanthus roseus*, although still in the family Apocynaceaè—a small, short-lived shrub with white or bright pink flowers. It sprang to international prominence in the 1960s when it was discovered that alkaloids in its sap had anti-cancer properties. Its chemical extracts, including vinblastine and vincristine, have become important pharmaceutical drugs, being especially successful against childhood leukaemias. The Madagascar periwinkle is now cultivated in fields across the tropics.

SHADY SPREADER
The greater periwinkle,
Vinca major (above),
produces soft blue flowers
in early spring. Its long,
wiry stems root when they
touch the ground and
colonies soon cover a wide
area, making it a good
groundcover plant.

PRETTY CANCER CURE
Catharanthus roseus, the
rosy periwinkle (right), is
a native of Madagascar,
but now found throughout
the tropics. In addition to
being an attractive plant
for cultivation, its sap
contains anti-cancer
properties.

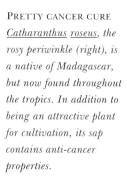

RANGING FROM THE BEAUTIFUL TO THE

BIZARRE, ORCHIDS ARE FOUND IN HABITATS AS

DIVERSE AS ARCTIC BOGS AND MANGROVE SWAMPS, BUT HAVE A PREFERENCE FOR TROPICAL CLIMATES.

Ophrys apifera, the bee orchid.

THE ORCHID FAMILY

(ORCHIDACEAE)

Orchids make up the second largest family of flowering plants after the Daisy family. Although there are orchids to be found in an extraordinary range of habitats, most of the 17,500 species grow in the tropics, with fewer than 2,000 found in temperate regions. Their name is derived from the Greek word *orchis*, meaning "testicles," and was prompted by the paired, round tubers of some European terrestrial species.

Orchid flowers are made up, in almost every case, of an outer ring of three sepals, and

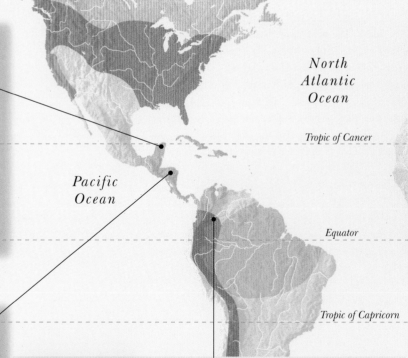

Stanhopea wardyi, *like all members of the genus, bears its large yellow and brown flowers* below the plant, necessitating its cultivation in a lattice pot slung from the greenhouse roof; in the wild they hang from the branches on which they grow. Natives of central and southern America, all stanhopeas need warm greenhouse conditions.

North Atlantic Ocean

Tropic of Cancer

Pacific Ocean

Equator

Tropic of Capricorn

Cattleya bowringiana *is one of the most free-flowering species in the genus. Rare in the wild, it grows on cliffs above streams in Guatemala and* Belize, from where it was introduced to Europe in 1884. Other species are epiphytes, but all should be grown in pots rather than baskets, using very well-drained, fibrous compost. *C. bowringiana* is named after a 19th-century English orchid grower, J.C. Bowring.

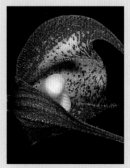

Dracula bella, *whose name means "beautiful little dragon," has nothing to do with vampires except a preference for shady conditions. Dracula has some 80 species in the Andes of northern South America, all of which have wonderfully grotesque* flowers in somber colors. Many are threatened in the wild by over-collecting, but even more seriously, by destruction of their forest habitat. In cultivation they need light shade and high humidity to flourish.

an inner ring of three petals surrounding the reproductive parts. Usually, two of the petals closely resemble the sepals, while the third forms a lip that is quite different in shape and, often, in color. The large, brightly colored flowers of *Cattleya* species and hybrids typify this standard pattern, but there are infinite variations, many of which are evolutionary adaptations which secure successful pollination for the flowers.

Pleione formosana
comes from the island of Taiwan—formerly known as Formosa—and mainland China, where it grows as an epiphyte, or among moss on rocks. Dormant in winter, it produces its flowers in early spring before the leaves emerge, and is easily cultivated in bright but cool conditions on a windowsill. It has been hybridized with species from western China to create a range of beautifully colored cultivars.

Spectacular cohabitants

Cattleyas are epiphytes, meaning that they grow on other plants but do not draw nourishment from them. Named for a pioneer of successful orchid cultivation, the English orchid collector, William Cattley (1788-1831), they come from the forests of tropical Central and South America. The species grown by Cattley was sent to him by a friend visiting Brazil as a non-flowering "filler" in a package of other plants. Later that year the unknown plants produced spectacular flowers, with pink petals and sepals, a rich purple lip and a yellow throat, which were 6 inches (15cm) across. The plant was named *Cattleya labiata* by the orchid expert John Lindley (1799-1865), but was not rediscovered in the wild for another 70 years. In the meantime, however, many other species of *Cattleya* had been introduced. One of these was *C. skinneri*, now the national flower of Costa Rica. This smaller species is bright rose-purple,

Key

Tundra

Boreal coniferous forest

Temperate woodland and grassland

Mediterranean

Desert

Tropical woodland and grassland

Tropical rain forest

A PREFERENCE FOR THE TROPICS

Indian Ocean

Pacific Ocean

Dendrobium spectabile *is just one of over 900 species of the genus found throughout Asia and Australiasia, spanning a great variety of habit, flower color, shape,and growing conditions. In the wild it usually grows on trees, but occasionally perches on rocks. D. spectabile comes from New Guinea and the Solomon Islands.*

Vanda Rothschildiana, *one of the bluest of orchids, is a hybrid between the epiphytic V. coerulea from Assam and Euanthe (Vanda) sanderiana from* the island of Mindanao in the Philippines. The paler blue *V. coerulea* was highly sought in the mid 1800s and was the subject of a very early plant protection law restricting its collection. *V. Rothschildiana* is the easier to grow of the two, but still requires a warm greenhouse to survive.

with a pale yellow throat. Its name honors George Ure Skinner, a merchant trader in Guatemala, who in 1831 was commissioned to collect orchids by the British amateur botanist James Bateman. Bateman (1811-97) had a lifelong interest in orchids; while a student at Magdalen College, Oxford, he had been made to copy out half the *Book of Psalms* for going to see an orchid flowering in the Botanic Garden across the road rather than attending a tutorial. Skinner sent his *Cattleya* to Bateman in time for it to be described in his monumental work, *The Orchidaceae of Mexico and Guatemala*, of 1838. This tome is one of the heaviest ever produced, each copy weighing 38 pounds (17kg).

Many orchid enthusiasts of those days hired professional collectors to supply them with plants. Competition was fierce, and collectors often deliberately attempted to gather every specimen of a rare species that they could find. Some even attached false labels of origin to their consignments to fool rivals. Enormous numbers of plants were collected, the great majority of which failed to reach their destination alive. Populations were devastated, and formerly common plants became rare overnight. In India, the blue-flowered *Vanda coerulea* became so rare that it had to be legally protected; while in Central America, collector Albert Millican chopped down so many trees in his hunt for orchids that the habitat of the pale pink *Miltoniopsis vexillaria* looked as if it had been cleared by forest fires.

A study in contrasts

The only "useful" member of the orchid family is the tropical American *Vanilla planifolia*. The vanilla essence extracted from the pods was used by the Aztecs to flavor chocolate, and the Spanish conquistadors introduced the pods to Europe in the 1500s. Plants followed in the early 1700s and the European powers, believing that vanilla would make a useful crop in their tropical colonies, distributed plants to many parts of the Old World. However, the absence of a suitable pollinator meant that no pods formed. It was not until 1841 that a former slave, Edmond Albius, living on the Indian Ocean island of Reunion, discovered a means of artificial pollination that meant

USEFUL ORCHID
Vanilla flavoring is derived from the seed capsules of the South American orchid Vanilla planifolia (above). The flowers are pale green.

COMPLEX HYBRID
The richly colored flowers of x Vuylstekeara (right) are a result of complex hybridization by a Belgian nurseryman. The hybrid thrives on a windowsill in a warm room.

vanilla could be produced throughout the tropics. Forming a robust vine, *V. planifolia* bears attractive, pale green flowers and is often grown as an ornamental.

Providing a complete contrast to *Vanilla* is the small genus *Dracula*. Lurking, like its vampirous namesake, in the shadows of the rainforest in the lower valleys of the Andes of Colombia and Ecuador, *Dracula* has flowers that are bizarre even for an orchid. The three sepals are fused to form a single, hoodlike structure which surrounds the reduced lip and petals. These resemble a distorted face peering out of the depths of the flower. The sepals are usually shades of dull red and orange and may be hairy or warty. The dull yellow flowers, striped with dark brown, are 12 inches (30cm) long, providing a striking feature for those who can provide the cool, extremely humid conditions it needs. *D vampira* was discovered in Ecuador, in the late 1800s, by the German Consul, F.C. Lehmann. However, the unfortunate Lehmann was drowned shortly after making his discovery, and his specimen lay unrecognized at the Royal Botanic Gardens, Kew until 1978.

Hybrids and hothouse beauties

Orchids have many elaborate adaptations to invite specific insect pollinators, but are easily hybridized by artificial means, and hybrids involving up to six genera have been recorded. By convention such complex hybrid groups are named for their raiser. Thus, in 1910, when the Belgian nurseryman Charles Vuylsteke released a complex hybrid involving the Neotropical genera *Miltoniopsis*, *Odontoglossum*, and *Cochlioda*, it was named x *Vuylstekeara*. The flowers of this group of orchids are flat, earning them the common name of "pansy orchids," and are usually pink or red. They grow well in centrally heated homes and have become popular house plants, with more than a million being sold annually in Europe alone.

Central heating has made it possible to grow many orchids successfully in the home, provided that you do not allow them to become desiccated. The moth orchids, *Phalaenopsis*, with their rounded white or pink flowers, are particularly tolerant, and will flower regularly in

MOTH ORCHIDS

<u>Phalaeonopsis</u> species (above), known as "moth orchids," are native to the forests of southeast Asia where they perch on tree branches. They have become widely cultivated as houseplants.

rooms where the temperature does not fall below 60°F (16°C). The earliest species known to science was *P. amabilis*. First described in 1750, it comes from Amboina, a small island in Indonesia. It has white flowers, as does *P. aphrodite*, a Philippine species, which bore the first *Phalaenopsis* blooms in Europe in 1839. Pink- and purple-flowered species such as *P. sanderiana*, were discovered later in the 19th century and became the foundation material for modern hybrids. Although *Phalaenopsis* is epiphytic, it is usually grown in pots. It is

essential to use a very course potting medium to ensure that drainage is perfect. Other orchids grow satisfactorily when attached to a piece of bark suspended from the roof, so long as the humidity remains high and the plants receive a regular spray of water.

Subtropical houseplants

Like *Phalaenopsis*, all *Cymbidium* species are natives of southeast Asia and northern Australia but, unlike the moth orchids, they are terrestrial, high altitude plants which can tolerate much cooler temperatures. This comparative hardiness enables hybrid cymbidiums to be widely planted in subtropical and warm temperate gardens, where temperatures do not drop below 40°F (5°C), as well as in greenhouses and conservatories in colder climates. The inflorescences emerge from the large clumps of stiff, erect leaves in spring and bear many long-lasting flowers in shades ranging from green to deep red, usually with the lip heavily marked in a contrasting color. The diminutive *C. ensifolium* has been cultivated in China for centuries. It is valued for its small but intricately marked flowers in shades of green and brown. The flowers are less than 2 inches (5cm) wide and are often scented. Cultivars with variegated leaves are highly prized. The first hybrid *Cymbidium*, a cross between the Asian species *C. eburneum* and *C. lowianum*, flowered in England in 1889; when crossed with *C. insigne* it gave rise to the hybrid group *C. Alexanderi*. The clone *C. Alexanderi* 'Westonbirt', its pale pink flowers spotted with red, first flowered in 1911. It went on to became one of the most important parents of the large hybrids that have dominated the orchid scene for most of the century. Recently, however, breeders have crossed these with smaller, wild species, such as *C. floribundum* from southwest China, to create miniature hybrids. The compact habit of these new hybrids makes them convenient houseplants.

Another popular group of indoor orchid are the pleiones, which bear large flowers at a height of no more than 3 inches (8cm). Named after Pleione, the mother of the Pleiades, who in Greek mythology became a cluster of stars, pleiones are found in the montane forests of the Himalayas and China, at altitudes up to 10,000 feet (3,000m), growing either in loose, humus-rich soil or among the moss on the trunks of rhododendrons and other trees. The flowers are a typical orchid shape, usually pink, their fringed lips spotted with yellow and brown inside. Many hybrids with more intense colors have been raised. Some, such as the Shantung group, even have yellow flowers. Most are easily grown on a cool window ledge. They should be planted in an open medium, watered heavily in summer and kept dry in winter.

Collecting mania

Many orchids are highly endangered in their natural environments, often because of habitat destruction, but also, sadly, as a result of over-

The birth of an orchid

Following successful pollination, an orchid flower produces a capsule containing thousands of minuscule seeds, each an almost invisible speck that is easily carried on the wind. The seed contains only an embryo and lacks the stored nutrients that would permit it to grow. On landing in a suitable place, it is invaded by a fungal strand from which the embryo absorbs nutrients. It begins to develop, and a minute "protocorm" appears. At first, this is totally dependent on the fungus for sustenance, but becomes less so as it matures and grows leaves. The fungus benefits from the "home" the mature orchid plant provides, making this a mutually supportive, or symbiotic association.

A few orchids remain dependent on the fungus throughout their lives and never produce chlorophyll (normally essential for food production in a plant). This mycorrhizal (from the Greek for fungus and root) association makes it almost impossible to grow orchids by sowing seeds in the conventional way. Instead they must be sown on a jelly rich in nutrients that can be absorbed by the embryo, which then develops as it would if a fungus were present. This remains a laboratory process requiring sterile conditions, but in skilled hands it is a reliable technique for producing healthy seedlings, which are eventually weaned and potted up conventionally. A similar technique, whereby a meristem (growing point) is cultured on nutrient jelly, is used to produce large numbers of plants, especially when propagating a particularly fine clone. Under ordinary conditions, orchids grow too slowly to provide more than a few offsets for propagation each year, but meristematic propagation can produce many hundreds of young plants from a small quantity of tissue. This is important if large numbers of identical plants are needed—for cut flower production, for example.

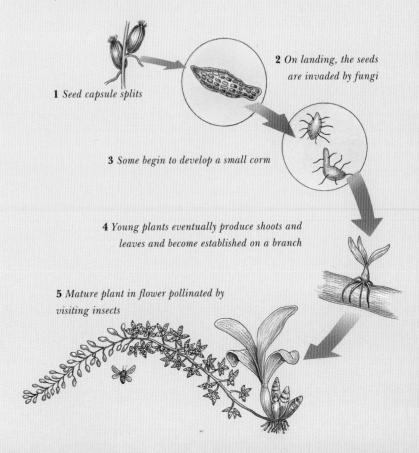

1 *Seed capsule splits*

2 *On landing, the seeds are invaded by fungi*

3 *Some begin to develop a small corm*

4 *Young plants eventually produce shoots and leaves and become established on a branch*

5 *Mature plant in flower pollinated by visiting insects*

*M*ELBOURNE BOTANICAL GARDENS
WERE FOUNDED TO CULTIVATE
INDIGENOUS AND EXOTIC PLANTS; TODAY THEY PLAY A VITAL ROLE IN
TEACHING THE PRINCIPLES OF PLANT ECOLOGY AND CONSERVATION.

*A*USTRALIAN *E*DUCATION

HOLISTIC APPROACH
Melbourne Botanical Gardens (above and right) combine the principles of ecological planting with a broad range of plants from all over the world.

Trom the 16th century onward, European powers became interested in the plants of their new overseas colonies. Botanical Gardens were established to investigate the commercial potential of the native flora, or to acclimatize new introductions. One of these was the Melbourne Botanic Garden in Australia, founded in 1846 as a "public domain for the purpose of rearing and cultivating indigenous and exotic plants" on ground sloping towards the Yarra river.

The Melbourne Botanic Garden was fortunate to have two remarkable early directors. Baron Ferdinand von Mueller (1825-1896), appointed in 1857, laid the foundations of a great botanical collection, both living and dried. But, although he was a first-class botanist, von Mueller had little regard for horticultural aesthetics. The garden was laid out in traditional, rectangular, systematic beds, and the trees were set in stiff plantations that a contemporary described as

"Prussian." Public reaction was unfavorable so, in 1873, he was replaced by William Guilfoyle, who immediately redesigned the garden as a landscape in which individual plants were less important than the overall effect. The result has been described as the best example of landscaping in Australia.

International collection
Today the Melbourne Botanic Garden cultivates some 12,000 species of plants from around the world. In fall, the oaks (*Quercus*) and elms (*Ulmus*), descended from the early introductions of economic plants, provide a strong contrast in color to the fine collection of gum trees (*Eucalyptus*). Elsewhere a gully has been planted with tree ferns and other lush vegetation in imitation of an Australian subtropical rainforest. Overhead a colony of flying foxes has found a roosting site, and on the lake are indigenous black swans.

All public gardens have a major role to play in educating the gardening public, introducing them to new ideas in horticulture, the plant sciences, and conservation. Melbourne takes this role very seriously, with education programs for schoolchildren and their teachers, as well as the general public. The Water Conservation Garden, for example, shows that drought-tolerant plants can be used to create a garden that is as attractive as one full of thirsty European perennials, still the most commonly grown garden plants in Australia. Water conservation is an increasing concern throughout the world, and the trial and recommendation of suitable plants for drier gardens is an ideal task for the expertise and resources of Melbourne and other botanic gardens around the world.

TOP LANDSCAPING
The gardens slope down toward the Yarra River, large lawns giving ample vistas to the river and other water features. The botanical collections fitted around this basic scheme.

1. *Clematis Shelter*
2. *Rose Garden*
3. *Australian Border*
4. *bulbs*
5. *Herb Garden*
6. *Fern Gully*
7. *cacti and succulents*
8. *Rock Garden*
9. *National Herbarium*
10. *Nymphaea Lily Lake*

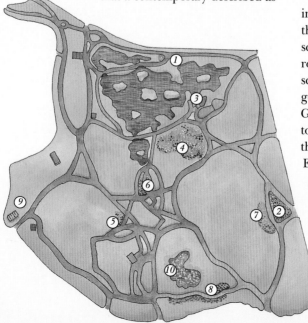

collecting for cultivation. In Britain, for example, the beautiful yellow and purple Lady's Slipper orchid, *Cypripedium calceolus* is a victim of both habitat loss and over-collection; for much of this century only a single wild plant survived in a secret locality, where it was constantly guarded from unscrupulous collectors. Modern propagation techniques have now made it possible to grow seedlings from this plant which are being planted in suitable locations near their wild parents.

The pouch-shaped lip of *C. calceolus* suggested a shoe, slipper, or clog to early European botanists, as reflected in its many names: *Frauenschuh* (women's shoe) in German, *Fruesko* (Virgin's shoe) in Norwegian, *sabot-de-Venus* (Venus' clog) in French. The English name is an abbreviation of "Our Lady's Slipper," itself a translation of the Latin *Calceolus Mariae*. Found in mountainous areas of central and northern Europe, *C. calceolus* and related species from Asia and North America make spectacular plants for a cool, moist site in a rock garden.

The closely related tropical genera *Paphiopedilum* and *Phragmipedium* are also known as slipper orchids and widely cultivated in greenhouses for their magnificent pouched flowers, often with long, twisted petals. New species introduced to Europe during the 1800s were cultivated with great enthusiasm. In fact, for many years, several species of *Paphiopedilum* were known only in cultivation, their wild origin being unclear. One of these was *P. fairrieanum*, named for a Mr. Fairrie, in whose orchid collection at Liverpool, England, it flowered for the first time in 1857. He had acquired it from an auction of imported plants, but its history was otherwise unknown. It became known as the "lost orchid" and was not rediscovered until 1904, when a surveying team found it in the foothills of Bhutan.

From neighboring Assam came *Paphiopedilum spicerianum*, a green, white, and purple flowered species that was to become an important parent of many hybrids. Its exhibition in 1878 caused a genteel furore among orchid growers, and it was traced back to a Mrs. Spicer, whose son, a tea-planter, had sent it home in a consignment of orchids. The interest shown prompted an enterprising nursery to send the German collector Ignatz Forstermann (1854-95) to collect plants from the wild. This he did, having to shoot a tiger in the process, but to forestall competitors he

arranged for the destruction of all the plants he could not transport back to Europe.

Ruthless collection continues to this day. In 1981 the magnificent bright-red *Phragmipedium besseae* was discovered in the forests of Ecuador by Mrs. Libby Besse. Within a few years the only known localities had been stripped of plants by collectors. To make matters worse, the forests in which it grows are now being logged, thus destroying its habitat forever. Fortunately *P. besseae* is easily grown in warm, humid conditions, and will produce a constant succession of flowers.

Luring a pollinator

The pouch-shaped lip of the slipper orchids is an adaptation to attract a suitable pollinator to the flower but, as with many orchids, it is a deceitful bribe, as no reward in the form of nectar or perfume is offered. As the insect pushes its way out of the flower the sticky pollen becomes attached to its head, hopefully to be transferred to the stigma of the next flower it visits. A similar shape of flower is found in *Stanhopea*, a tropical American genus of epiphytes. In this case, the flowers are borne beneath the plant, offering easy access to the bees that visit them. The attractant is the powerful fragrance, which is collected by the bees as they crawl through the flower and used by them in courtship rituals, each species of orchid attracting a different species of bee. The pendulous inflorescence means that it is essential to grow stanhopeas in baskets suspended from the greenhouse roof. The Mexican *S. tigrina* is one of the most commonly cultivated species; its large yellow and brown flowers may be 8 inches (20cm) across.

Some orchids, especially those pollinated by moths, offer a more tangible reward. The classic example is the Madagascan orchid, *Angraecum sesquipedale*, whose large white flowers have a nectar-filled spur 12 inches (30cm) or more long. Discovered in the early 1800s, it aroused the curiosity of Charles Darwin (1809-82), who recognized in the adaptations of orchids to their pollinators a fine justification of his theory of evolution by natural selection. The pollinator of *A. sesquipedale* was unknown to Darwin, but he predicted that a moth would be found with a proboscis long enough to reach the base of the spur. His prediction was proved correct, and the large hawk-moth concerned was named *Xanthopan morganii praedicta*.

LOST AND FOUND
The wild home of
Paphiopedilum
fairrieanum (right)
remained unknown for
nearly 50 years after it
flowered in an English
greenhouse. It was
eventually found in a
remote part of Bhutan in
the Indian subcontinent.

*F*AMILIES OF SILKY-PETALED

FLOWERS FROM THE SUNNY COASTS

OF CALIFORNIA, FARMLANDS OF EUROPE, AND ORIENTAL FORESTS

*P*OPPIES AND *P*EONIES

(PAPAVERACEAE AND PAEONIACEAE)

A WORLD OF POPPIES
the main genera and their origins

**Eshscholzia
californica**

*The California poppy—
the State flower—is an
annual, or short-lived
perennial, germinating
during winter rains and*

*flowering in spring. The
flowers emerge from an
undivided pointed calyx
which looks like a pixie's
cap as it is forced off by
the opening petals, and
the flowers are followed by
an elongated capsule
which explosively expels
the ripe seeds.*

Argemone mexicana
*The stems, leaves, and
calyx of the Mexican or
prickly poppy are covered
in sharp spines. It is
native to Central America
and the Caribbean*

*Islands—another common
name is the Jamaican
thistle. This species bears
pale yellow flowers over
glaucous blue foliage;
other species have white
flowers. All are most
satisfactorily cultivated as
half hardy annuals.*

Key

Tundra	Mediterranean
Boreal coniferous forest	Desert
Temperate woodland and grassland	Tropical woodland and grassland
	Tropical rain forest

Meconopsis cambrica

The Welsh poppy takes its name from the hills of Wales, or Cambria, where it is common in woodland, and grows in similar habitats elsewhere in western Europe. It is the only Meconopsis to occur in Europe; the others are all natives of the Himalaya and China. In cultivation it is an attractive perennial for cool, shady spots, and also tends to establish itself in old walls. Like many poppies, it produces abundant seeds and can become weedy. To prevent this becoming a problem, remove the capsules before the seed is shed. Though typically bright yellow, orange and brick red forms are known, and doubles occur in all shades.

Poppies are generally very adaptable to different environments. They grow and reproduce very quickly and can make perfect candidates for weed status. Each seed capsule produces hundreds of seeds, capable of great longevity in the soil, and which germinate only when exposed to light.

Poppies thrive in disturbed ground. The scarlet common corn poppy, *Papaver rhoeas*, became a symbol of remembrance for the war dead after it appeared in moving masses on the war-torn soil of northern France. It also thrives on farmland and other cultivated ground. In

Papaver somniferum

The opium poppy is an easily cultivated annual which enjoys well drained soil in full sun. The flowers are typically pale pink, but gardeners have selected cultivars with shades varying from white to black-purple, often double or with fringed petals— and with diminished narcotic powers.

1880 the Reverend W. Wilks (1843—1923), vicar of the parish of Shirley in Surrey, England and later Secretary of the Royal Horticultural Society, noticed that among a mass of ordinary scarlet poppies in an odd corner of his garden, was one plant with flowers narrowly edged with white. He marked the plant, and sowed its seed, weeding out all the plain seedlings when they flowered, and saving seed from the most distinct. By means of this diligent selection, continuing over many years, Wilks created the famous strain of Shirley Poppies, from which all trace of the scarlet and black of their ancestors was removed to create orange, pink, and white versions, often with a picotee margin. Another strain, also raised in England, are the 'Mother of Pearl' poppies selected by the artist Sir Cedric Morris (1889—1982), which flower in an ethereal haze of lilac, gray, pink, and white, and are also known as 'Fairy Wings.'

Bringer of sleep and forgetfulness
Other *Papaver* species have for centuries been associated with sleep and forgetfulness. While *P. rhoeas* is merely a weed to the farmers of temperate climes, Burmese farmers would be delighted to see a field of *P. somniferum*, the opium poppy, bringer of sleep. It is an easily grown annual and one of the oldest of cultivated plants. The Sumerians grew crops of opium poppies in the "fertile crescent" of Mesopotamia between the Tigris and Euphrates rivers from 5,000 B.C. both for food and the narcotic, sleep-

REVERENTIAL SELECTION
An English clergyman noticed a wild scarlet poppy with white-edged petals in a corner of his garden, and began a long selection process which produced Shirley poppies. They come in a variety of pale shades and have lost the dark center of their wild ancestors.

THE CORN POPPY
Papaver rhoeas (far left) flourishes in disturbed soil, whether on a battlefield or a neglected corner of farmland.

Betty Sherriff's dream poppy

The popularity of *Meconopsis betonicifolia* started something of a craze for blue poppies, especially in Scotland, where they enjoy the cool, moist conditions. Of 43 members of the genus *Meconopsis* (as currently recognized), only one, *M. cambrica*, the Welsh Poppy, occurs in Europe; the others are all from the Himalaya and China, where there are species with white, yellow, red, or purple flowers as well as blue.

In 1933 the Scottish plant hunters George and Betty Sherriff were in Bhutan. One night in May, when George Sherriff was elsewhere, Betty had a vivid dream, in which he appeared and gave her clear directions to a site for a new poppy, even seeing her husband wag his finger at her and say "Be sure you go." Next morning, despite her sceptical companions, Betty Sherriff followed the vision's directions, which were absolutely accurate and led her to a glorious blue poppy, a form of *M. grandis*. Seeds were collected, and their descendants are still cultivated as "Betty's Dream Poppy."

I myself can confirm the thrill of finding a blue poppy. When in 1988, at the age of 20, I set off for Nepal for a summer of plant hunting, I had one burning ambition -to see a blue poppy in the wild. That ambition was achieved on a day of dank mist, through which shone the unforgettable, intense bright blue of *Meconopsis horridula*, set off by the bunch of golden anthers inside the flower. It was the most satisfying sight imaginable, never equaled before or since.

HIMALAYAN BLUES
*Meconopsis betonicifolia is
the most easily grown
Himalayan blue poppy. It
is a perennial which
flourishes in cool, moist
climates like those of its
native mountains in
southern Tibet. The
astonishing color develops
to perfection only in acidic
soil, preferably enriched
with organic matter.*

inducing properties of the sap. Opium was used as a painkiller by the ancient Greeks from 800 B.C., and remained the only effective analgesic until the advent of modern medicine. Even today, morphine and other compounds are extracted from poppy sap by the pharmaceutical industry.

Although *P. somniferum* is legitimately cultivated in many parts of the world, an enormous acreage is harvested for illegal narcotics, especially heroin (the refined form of opium). This is nothing new; much of the wealth and influence of the British Empire was founded on the opium trade.

The flowers of the opium-bearing poppy are a dull, pale mauve or white, but cultivated varieties (which have a much lower alkaloid, and thus narcotic content) may be white to pink, scarlet, or blackish-purple, often doubled to enormous size, or with fringed margins. They have been cultivated in the West since the 1500s, and often featured in 17th- and 18th-century Dutch flower paintings.

Blue wonder
In 1927, the superintendents of Hyde Park, London, and Ibrox Park, Glasgow, planted a blue "poppy" which had been newly introduced from Tibet in some of their flower beds. It was a sensation. Botanists and keen gardeners had been aware of the existence of a wonderful blue

poppy for years, but no specimen had been a success in cultivation. The poppy was *Meconopsis betonicifolia*, first collected in 1886 by the French missionary Pére Delavay. It remained a herbarium specimen, however, until, in 1913 the English explorer F.M. Bailey, traveling through southern Tibet, found and pressed a single blue poppy flower. The plant hunter Frank Kingdon-Ward (1885—1958), traced Bailey's steps in 1924 and refound the poppy in full flower:

"Suddenly I looked up and there, like a blue panel dropped from Heaven, a clump of blue poppies—as dazzling as sapphires—gleamed in the pale light." Kingdon-Ward immediately recognized the flower's potential as a cultivated plant and returned later in the year to collect seed from which grew the plants which astonished the public and press in 1927. From there, meconopsis became firmly established all over the world. It grows naturally in woodlands in the cool, moist climate of the eastern Himalaya and enjoys moist, fertile conditions in cultivation. In warm areas it needs as cool a site as possible. The usual, purest blue of its flowers can be affected by soil conditions; dry or limy soil may cause the flowers to be tinged purple.

Flower of the golden west
In spring, after a wet winter, the hills of California become a golden sheet of poppies—*Eschscholzia californica*—which has become the State flower. It was once dedicated by the Spanish colonists of Central America to San Pascual, but its formal, Latin name—one of the most commonly misspelled in horticulture—honors the German Johann Friedrich Elsholz, who took part in a Russian expedition to California in 1815, and died there at the age of 38. The botanist Ludolf Chamisso (1781—1838), who described the plant, couldn't spell either, and called him Eschscholz. The California poppy is an annual, usually germinating in late fall or winter, and flowering in the spring before being scorched by summer heat. If the Californian winter is dry, the display of poppies and other flowers is much less spectacular. In temperate climates, *Eschscholzia* does best when grown as an annual in sunny, dry places; in too rich or moist a soil it becomes very leafy and less floriferous.

Other poppies found in California include the sumptuous *Romneya coulteri*, a shrubby plant with glaucous leaves that bears huge white flowers with a central boss of golden anthers which have been rudely compared to a fried egg. *Romneya* grows in the chaparral, the dense, rather scratchy and aromatic scrubland that covers the hills of southern California, where it spreads by stolons to form large colonies. It dislikes being moved, so try to establish a young plant in a warm, well-drained location and leave it undisturbed. The name commemorates two Irishmen, the Reverend Romney Robinson (1792—1882), an astronomer from County Armagh, and the actual discoverer, Thomas Coulter (1793—1843). Coulter was a doctor who also managed mines in California, and in his free time made many important discoveries in the Californian flora.

Prickly relations

The prickly poppies, *Argemone*, have large, white and yellow flowers similar to those of *Romneya*, and are native to the southwestern U.S. and western Central and South America. Species

TISSUE PAPER FLOWERS
The delicate flowers of Romneya coulteri, the matilija, or tree poppy, can be seen from the freeways of southern California, like leaves of tissue paper among the scrub. In cultivation they enjoy sandy soil in full sun.

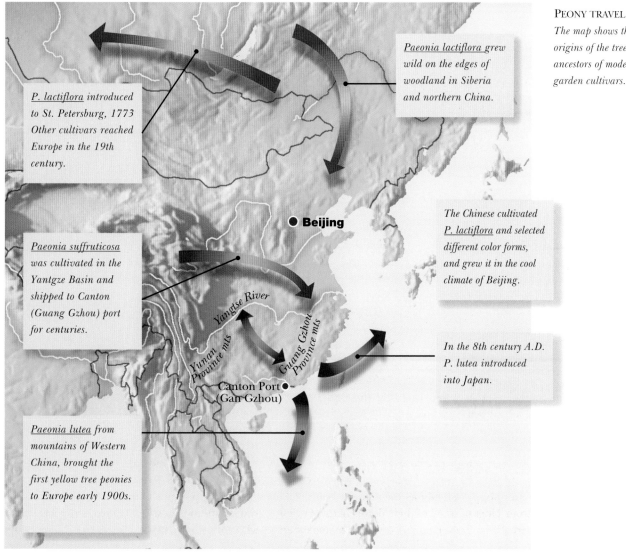

PEONY TRAVEL
The map shows the wild origins of the tree peonies, ancestors of modern garden cultivars.

Paeonia lactiflora grew wild on the edges of woodland in Siberia and northern China.

P. lactiflora introduced to St. Petersburg, 1773 Other cultivars reached Europe in the 19th century.

The Chinese cultivated P. lactiflora and selected different color forms, and grew it in the cool climate of Beijing.

● **Beijing**

Paeonia suffruticosa was cultivated in the Yantgze Basin and shipped to Canton (Guang Gzhou) port for centuries.

Yangtse River

Yunan Province mts

Guang Gzhou Province mts

Canton Port ● (Gan Gzhou)

In the 8th century A.D. P. lutea introduced into Japan.

Paeonia lutea from mountains of Western China, brought the first yellow tree peonies to Europe early 1900s.

$\mathcal{T}$HE DRY HILLS ABOVE LOS
ANGELES ARE THE UNLIKELY
SETTING FOR THE OPULENT HUNTINGTON BOTANICAL GARDENS, WHERE
HORTICULTURAL STYLES FROM AROUND THE WORLD HAVE BEEN
HARMONIOUSLY UNITED.

$\mathcal{M}$ILLIONAIRE'S $\mathcal{G}$ARDEN

The 600-acre (242ha) estate was a farm when bought in 1903 by Henry Huntington, who had made an enormous fortune in the real estate business. In the hills of southern California he built a mansion for himself and his art collection, a library for his rare books, and gardens which drew inspiration from a range of gardening traditions.

One of Huntington's priorities was the establishment of a palm garden incorporating grand vistas and Italianate statuary. Given the local weather with its hot summers and severe winter frosts, this was a daunting task. Eventually, enough palms survived to provide shelter for younger plants, and the Palm Garden now grows 200 species in 80 genera, many of which are rare elsewhere in cultivation. Its success was largely due to the exploitation of microclimates, one of the secrets of successful gardening. Easier to manage are the water gardens, where a collection of temperate and tropical water lilies coexist with red-eared terrapins. Nearby are collections of bamboos and unusual conifers, including a specimen of *Taiwania cryptomerioides*, the "coffin tree," whose wood is used in China to make coffins.

The Huntington gardens unashamedly rely on extensive irrigation to grow most of their plants, despite the drought conditions often prevailing in southern California. This extravagance permits the cultivation of a fine collection of camellias and a rose garden with cultivars arranged in historical order. Other notable features include a "Shakespeare garden" inspired by the Shakespeare "First Folio" in the Huntington Library. Here an English cottage garden has been recreated, containing many plants which would have been familiar to Shakespeare himself. The Desert Garden does not attempt to recreate a desert

1 Desert garden
2 Palm garden
3 Jungle garden
4 Lily ponds
5 Subtropical garden
6 Australian garden
7 Rose garden
8 Herb garden
9 Shakespeare garden
10 Camellia garden
11 Camellia garden
12 Japanese garden
13 Zen garden
14 Orange grove

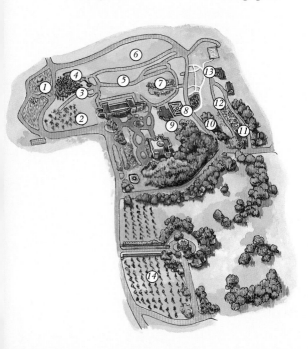

ecosystem, but is a landscaped collection of drought-tolerant succulent plants from around the world, positioned according to their geographical origins and botanical affinities. Henry Huntington was reluctant to include the desert garden, having once had an unfortunate experience with a cactus in the Arizona desert, but it is now one of the Huntington's most famous features, and a focus for its conservation work. Over 4,000 species grow in the 12 acres

(4.5ha), with another 2,000 in a nearby conservatory. The origins of each specimen are carefully documented, an important task if material from the collection were to be used for a reintroduction program.

Closely associated with the Huntington's Desert Garden is the International Succulent Introductions Program, which makes artificially propagated succulents available to enthusiasts throughout the world.

RECORD COLLECTION
The Desert Garden has the largest outdoor collection of desert plants in the world.

such as *A. platyceras* are easily grown in sunny, well-drained conditions, where they freely produce their magnificent flowers.

PEONIES: WHERE EAST MEETS WEST

Peonies, found throughout Europe and Asia, with a handful of species in western North America, are appreciated as cultivated plants equally in both the West and the East. The Greeks named the plant after Paeon, god of healing, who reputedly used the roots to cure a wound inflicted by mythical strongman Hercules on Pluto, god of the Underworld. Mortal Greeks were generally wary of the plant, believing that if it was dug up in daylight, woodpeckers would peck out the digger's eyes, or, like the medieval tales about the mandrake (*Mandragora* spp.), that it had to be pulled up by a dog, as its groan on leaving the ground was fatal to those who heard it.

In Europe, all peonies are herbaceous plants, mostly occurring in Mediterranean regions. Some, such as *Paeonia mascula*—the "male" peony—were cultivated in monastery gardens elsewhere for their medicinal properties. *P. mascula* is now seldom seen in gardens as its position was usurped in the 1500s by *P. officinalis*, the so-called "female" peony. A double form, 'Rubra Plena,' with huge, fully double, crimson flowers arrived in England from the Netherlands in the mid 1500s, when it commanded enormous prices. Its hardy vigor and magnificent flowers has earned it the Royal Horticultural Society's Award of Garden Merit, a distinction bestowed only on the very finest cultivated plants. All peonies flourish in open conditions in fertile soil, and should not be disturbed once established.

Doubling up

During the 1800s, many clones of *P. lactiflora* were imported to Europe from China, and crossed with *P. officinalis* to create a new strain of border peonies. A native of Siberia, from where a wild form with single white flowers was introduced to Europe in 1784, *P. lactiflora* had been grown and admired in China for a thousand years, and many fine double cultivars had been selected. The doubling is caused by the conversion of anther filaments into petals,

HERBACEOUS PEONIES
A stalwarts of the early summer herbaceous border since medieval times, Paeonia officinalis can survive neglect and still flower well each year, though it resents disturbance of its roots once established.

UNPRONOUNCABLE
Paeonia mlokosewitschii is a Caucasian species more easily referred to in English as "Mollie-the-Witch." The flowers last only a few days, but are followed by red and black seeds, and the red young shoots and grayish-green foliage are also handsome.

or sometimes takes the form of a thickening of the filament into a quill-like structure. The cultivar 'Bowl of Beauty,' with pink petals and yellowish-white quills, typifies this group. These double forms are not only white, but pink and red, and during the late 19th and early 20th centuries were hybridized with European peonies, both in Europe and America, to create the familiar border peonies with huge, double flowers. *P. lactiflora* is strongly scented, a character shared by many of its hybrid progeny.

Ever since its introduction to Europe from a limited area of the Caucasus in 1907, *P. mlokosewitschii* has challenged the vocal chords, and has earned the easier English name of "Mollie-the-Witch." It is one of the few yellow herbaceous peonies, extremely handsome and hardy. It emerges in spring with hairy, purple-tinged foliage, from which rounded flowers open in late spring. The flowering season is very brief, but if the flowers are successfully pollinated, they are followed by capsules of seeds which split open in late summer to reveal rows of shiny black seeds, separated by infertile bright red seeds.

King of Flowers

The shrubby, "tree peonies" derived from *P. suffruticosa*, were popular cultivated plants in China as early as the 6th century A.D., and were grown at first for their medicinal properties. The Chinese call *P. suffruticosa* the "king of flowers," and often portray it with the phoenix, king of birds. Its Chinese name is *moutan*— "most beautiful"—and the finest and rarest cultivars changed hands for vast sums. The plants were closely guarded in the grounds of the emperor and his mandarins, and *P. suffruticosa* did not reach Europe until 1787, when Sir Joseph Banks procured a specimen for Kew Gardens in London.

The Chinese practice was to grow the plants in nurseries far inland, and to ship small plants, while dormant, downstream to coastal towns. There they were potted and sold as they came into flower, soon dying in the heat of the coast. This routine meant that it was very difficult for Europeans to acquire healthy plants which would stand the long journey to Europe. Although many magnificent clones with white, pink, and red flowers did reach Europe during the 1800s, the true wild form remained

unknown, both to the West and to the Chinese. A manuscript of 1698 told of a hill in Shensi Province which was colored red with the flowers, but no peonies were found there when William Purdom visited it in 1912. It is possible that they had been exterminated by the collection of their roots for traditional medicine, a practice that is threatening the few remaining wild populations today. Purdom did find wild tree peonies in 1910, and in 1914 Reginald Farrer found hillsides covered with them in Gansu, being led to them by their "celestial fragrance," but no seed was collected. In about 1925—the precise date is uncertain—the American plant hunter

Joseph Rock (1884—1962) visited the monastery of Joni in Gansu, and found fruiting bushes of a tree peony, which the monks said had come from the nearby mountains. When seedlings flowered in cultivation it was found that here at last was a wild form of *P. suffruticosa*. With enormous white flowers, marked with crimson and maroon at the base, Joseph Rock's peony (now known as *P. rockii*) has become one of the legendary plants of horticulture. It has proved extremely difficult to propagate by grafting, the normal means of increasing tree peonies, and seed is rarely set; supply never meets demand.

AMERICAN DISCOVERY
Paeonia rockii, is often called "Joseph Rock's peony" after its American discoverer. It is the wild ancestor of the cultivated tree peony, *P. suffruticosa*, and originates in northwestern China. It is very hardy but the young shoots like protection from early morning sun.

PRIMULAS GROW NATURALLY IN
MOST TEMPERATE AREAS OF THE
NORTHERN HEMISPHERE, FROM THE MOUNTAINS OF AMERICA'S PACIFIC
COAST TO THE WOODLANDS OF EUROPE AND ASIA, WHILE CYCLAMEN ARE
NATIVES OF THE MEDITERRANEAN COUNTRIES.

Primula vulgaris

THE PRIMROSE FAMILY
(PRIMULACEAE)

Over the centuries, collectors have gathered wild *Primula* species from some of the most remote places on earth. *P. sikkimensis* was discovered among the mountain bogs and streams of Sikkim, Nepal, and Yunnan, where in late spring, the flowers tint the grassy banks a delicate yellow. The drumstick primrose, *P. denticulata* was found on open grassy slopes in Afghanistan and China. There are handsome native species in North America too, including the delicate, reddish-lavender Sierra primrose, and the brilliant, dark pink Parry's primrose, but the starting point for the creation of most of the multicolored strains of primrose and polyanthus grown today was the pale lemon yellow, European wild primrose, *P. vulgaris*.

The generic name from the Latin *primus*— "first"—indicates the early-flowering habit of

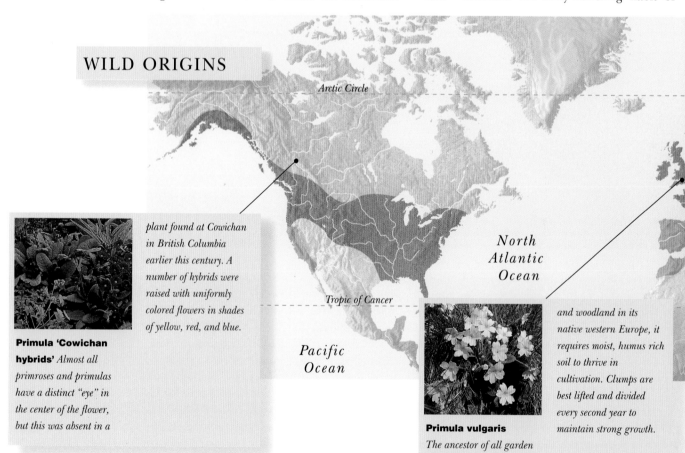

WILD ORIGINS

Arctic Circle

Tropic of Cancer

North Atlantic Ocean

Pacific Ocean

plant found at Cowichan in British Columbia earlier this century. A number of hybrids were raised with uniformly colored flowers in shades of yellow, red, and blue.

Primula 'Cowichan hybrids' *Almost all primroses and primulas have a distinct "eye" in the center of the flower, but this was absent in a*

Primula vulgaris
The ancestor of all garden primrose and polyanthus hybrids. Found in hedges

and woodland in its native western Europe, it requires moist, humus rich soil to thrive in cultivation. Clumps are best lifted and divided every second year to maintain strong growth.

many European species. The Italian *primaverola* is a diminutive of *fior di prima vera*, which means "the first flower of spring." This evolved into the French *primevere* and the English "primrose." While the European members of the genus are often early spring flowers, those that grow wild in the mountains of Asia and North America

Primula vulgaris subs. sibthorpii *is the counterpart of* P. vulgaris *found in the eastern Mediterranean area and Turkey. Its flowers are white or pink to purple,* *seldom yellow. The bright colors of modern hybrid primroses are mostly descended from purple forms. Like the western primroses, this subspecies is easily grown in a cool place but is seldom seen today as its place has been taken by more flamboyant descendants.*

flower as the snow melts, which may not be until late spring or early summer.

Introducing "Turkey red"

Although white and double forms of *P. vulgaris* were known in the 1500s, no northern European had seen a pink or red version. Then, in 1638, the English plantsman gardener John Tradescant the Younger collected what became known as the 'Turkey Red' while visiting Greece and Turkey. This plant, which was later officially named as *P. vulgaris* subsp. *sibthorpii* after John Sibthorp (1758-1796), the Englishman who laid the foundations of modern Greek botany, was to mark the beginning of colored primrose breeding. Merchants transported boxloads of plants to English and French gardens. Demand was so great that each rootstock was divided several times, and each section sold at an extortionate price.

The 'Turkey Red' is found from the eastern Mediterranean region east to the Caucasus mountains and varies considerably in the depth of color. Tradescant's original collection seems to have been a deep purple and was also prized

Primula sieboldii *is a native of northeastern Asia. In Japan it has been grown for centuries as a pot plant. It grows in fertile soil in partial shade, disappearing into dormancy in late summer, only to reappear the following spring.*

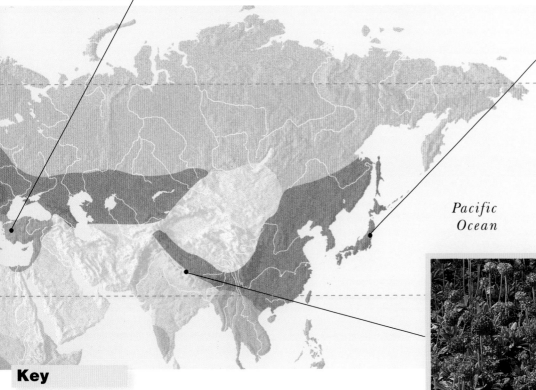

Pacific Ocean

Primula denticulata *is found in alpine grassland from Afghanistan to western* China *and may be extremely abundant in favored places. In rich soil it will soon make a large clump and may be propagated by seed or division. The plant's rounded inflorenscences have earned it the common name of "drumstick" primula.*

Key

Tundra	Mediterranean	Tropical rain forest
Boreal coniferous forest	Desert	
Temperate woodland and grassland	Tropical woodland and grassland	

STAR BACKING

Yellow polyanthus (above) were popularized by the great British plantswoman Gertrude Jekyll in the early 1900s.

HIMALAYAN COWSLIP

Elegant and deliciously perfumed Primula florindae *(right) is powdered with a white wax which may protect it from insect pests. It needs rich soil and ample moisture during the growing season to achieve full height.*

for its early flowering. Within a few years the Turkey Red had been crossed with the established forms and by 1648 purple, blue, white, and double white primroses were flourishing in the University Botanic Garden in Oxford, England. A double red appeared by the mid 1700s, but it became evident that only the purest hues could be used for selection, as hybrids between yellow and pink could turn out muddy in color. Double primroses are still grown, but are less vigorous than the single varieties, and need good moist soil and regular division. Although a blue was claimed in Oxford in the 1600s, it disappeared and a blue strain was not fixed until the late 1800s.

Color coordination

Another breakthrough came in the early 17th century with the creation of a red polyanthus hybrid, which capitalized on the colored forms of *P. vulgaris* combined with the long, strong stems, each with its umbel of small flowers, of the wild cowslip (*P. veris*), a European meadow plant. Selection for color and flower size progressed in parallel with the continuing development of the primrose.

The polyanthus, however, became a "florists flower" and was bred to a high standard of perfection by artisan gardeners, especially in the English counties of Lancashire and Cheshire. Their favored form was an almost rounded flower with a deep, velvety, black-red background, edged with bright yellow or white. These "gold-laced polyanthus," as they are known, are still cherished by gardeners, but are less vigorous than the bunch-flowered polyanthus. They enjoyed a revival of popularity through the English garden designer and

author Gertrude Jekyll (1843-1932), who developed a floriferous strain of yellow and white polyanthus.

The pianist and the polyanthus

Not all the brilliant colors of the modern primrose and polyanthus cultivars can be traced back to the 'Turkey Red'; some are hybrids between *P. vulgaris* and the Caucasian species *P. juliae*. A direct descendant of this cross is the bright purple cultivar 'Wanda' and the strain of seedlings raised from it. Another key contribution was made by an out-of-work American concert pianist named Florence Bellis early this century. She worked from a leaky timber cabin in Oregon, hand-pollinating by the light of an oil lamp and with only a wood stove for warmth. Bellis had no formal knowledge of plant genetics, but her Barnhaven primrose strains became internationally known, and in the 1940s, she produced a true blue variety. The Barnhaven and other strains were raised through direct selection of the best colors, but most retain the distinctive "eye" like those of the wild primrose and cowslip. One strain which does not is the 'Cowichan' polyanthus, which bears uniform, often deep-hued flowers, and is descended from a plant found in an old backyard in British Columbia.

Primulas from Asia

With few exceptions, alpine primulas from Europe and North America are plants for specialists only, as their cultivation requirements are not easily satisfied in an ordinary garden. Fortunately, many Asian species have proved amenable to cultivation, and grow especially well in cool, moist conditions. The first to be cultivated in Europe was *P. denticulata*, the drumstick primula, which bears tight pompons of mauve, white or red flowers in early spring. First introduced in 1838, it occurs over a range of nearly 2,000 miles (3,000km) in the Himalaya and associated ranges, from the Hindu Kush to western China, with a wide altitudinal range— and unsurprisingly is very adaptable. Drumstick primulas are easily grown in any soil that does not dry out in summer, and can make a useful spring bedding plant.

My own garden, in southern England, is too hot and dry for Asiatic primulas to flourish, but one with which I am prepared to struggle to keep going is *P. florindae*, the Tibetan cowslip, which I value particularly for its heavenly perfume and soft yellow flowers. When doing

Conservation issue

In 1901 the Dutch Bulb firm of Van Tubergen sent some cyclamen tubers, recently received from Turkey, to a Professor Hildebrand in Berlin for identification. When they flowered they turned out to be a new species, which Hildebrand, noting the red flush on the young leaves, named *Cyclamen mirabile*, meaning wonderful cyclamen.

Unfortunately, all the plants in cultivation died, and no more were imported. Then, the type specimen in the Berlin Herbarium, was a casualty of Allied bombing in the Second World War, leaving botanists with no material record. It was eventually rediscovered in the wild in 1956 and limited numbers were established in cultivation.

In the mid 1970s *C. mirabile* suddenly became very common. They were traced back to the horticulture departments of large chain stores, which had sold them as dry tubers. Investigation showed that massive exploitation of Turkish wild bulbs was going on, not only of cyclamen but also of snowdrops (especially *Galanthus elwesii*) and other small bulbs. At the peak of the trade in the mid 1980s some 6,600,000 cyclamen tubers were exported from Turkey annually. International concern led to the regulation of the trade by CITES, the Convention on International Trade in Endangered Species, which limits by quota the permissible annual harvest. Customer reaction against wild collected plants has also lessened the pressure, while village communities are encouraged to grow stocks of bulbs in nurseries as a sustainable source of income.

well *P. florindae* is a big plant, with stems reaching to over 3 feet (1m) in height and bearing great mops of flowers, making a very significant feature. The plantsman Frank Kingdon-Ward, who found it growing by (and even in) woodland streams in enormous numbers, in southeastern Tibet in 1924: "It choked up ditches and roofed the steepest mud slides with its great marsh marigold leaves; then in July came a forest of masts, which spilled out a shower of golden drops, till the tide of scent spread and filled the woodland and flowed into the meadow..."

His discovery, named after his wife Florinda, is best suited to a large bog garden, where it will self sow and maintain itself. It has become established in the wild in many parts of northern Europe, growing beside streams, just as it does in its land of origin.

Elegant candelabras

Another popular group of Asian primulas are the "candelabra" species and their hybrids, in which whorls of flowers are arranged in tiers up the flowering stem. With flowers in shades of orange and pink to bright red and magenta they are often seen in massed displays in waterside situations, although the effect can be one of rather barbaric splendor. The first to be introduced, and still one of the most widely

WATERSIDE ELEGANCE
Several Asiatic primulas bear their flowers in whorls or tiers up the stem, a habit which has earned them the name of candelabra primulas (above). Many thrive in waterside locations.

OPTIONAL FRINGE
The Japanese native Primula sieboldii *has given rise to many delicate cultivars in shades from white to deep purple, including some with fringed petals.*

grown, was *P. japonica*. Seeds and plants sent from Japan by Robert Fortune in 1861 failed to become established, but success was achieved in 1871. Exploration of western China yielded many more species, from which many strains of garden hybrids have been raised. For example, the bright scarlet-flowered *P.* 'Ravenglass Vermilion' (often called *P.* 'Inverewe' after the Scottish garden where it became well known), is derived from crosses between three Chinese species, *P. beesiana* (magenta), *P. bulleyana* (orange) and *P. cockburniana* (orange-red). Although sterile it has great hybrid vigor.

The cherry blossom primula

Each spring the citizens of Tokyo can visit the one remaining meadow where *P. sieboldii* grows wild. In the past they dug up a few roots and took them home, but now the species is very rare in Japan. Its local name is *sakurasoh*, from *sakura*, meaning cherry blossom, which the flowers resemble, and *soh*, meaning herbaceous plant. Japanese horticulturalists have selected an enormous range of varieties from *P. sieboldii*, including pinkish purple, white, deep red, and sometimes bicolored strains. The flower shape broadened and rounded as the petals overlapped, and the petals have become laciniate and feathery. The selections are grown in special pots and displayed on purpose built stands. As with other Japanese plants, the first *P. sieboldii* to be grown in Europe were cultivated selections rather than the wild original, introduced by Phillip von Siebold (1791-1866) to the Netherlands in about 1860. Von Siebold, an employee of the Dutch East India Company, sent many Japanese plants to Europe, where he grew them in a nursery at Leiden before distributing them more widely.

CYCLAMEN: "WITH LEVERET EARS LAID BACK"

The distinctive flowers of *Cyclamen*, with their upswept petals, described so evocatively by the English author and gardener Vita Sackville-West, may seem to share little resemblance to *Primula* species, but they share the basic flower shape of five "petals" united at the base. In North America, a similar flower shape and color is seen in *Dodecatheon*—the "Shooting Stars"—a member of the Primulaceae, although not closely related to cyclamen.

The large house plant forms of cyclamen are derived from the wild *C. persicum*, a native of the eastern Mediterranean region. It often grows in old olive groves: large numbers can be seen in spring under the ancient olive trees in the Garden of Gethsemane in Jerusalem. Although its scientific name suggests a Persian origin, it has never been found further east than Israel and Syria. In its wild form, *C. persicum* has white flowers tipped with pink at the "nose," narrow, elegantly twisted petals and a delicious perfume—one plant in full flower will scent a whole greenhouse. It has been grown in northern gardens since the 1600s, although, because it is frost-tender, never to any great extent. It became fashionable in the mid 1800s,

POTTED VERSION

Cyclamen hybrids used for winter houseplants were first bred in the 1800s from the much daintier wild Cyclamen persicum *from the eastern Mediterranean basin. They need cool, bright conditions and should be placed away from direct heat.*

North Sea

Danube

Black Sea

Euphrates & Tigris

Mediterranean Sea

Key

Cyclamen rohlfsianum

C. persicum

C. repandum

C. africanum

C. hederifolium Y

C. coum

CYCLAMEN SPECIALITIES
When the Mediterranean Sea flooded billions of years ago, fragmentation and specialization occurred among the Cyclamen genus, and each species evolved, for example, its own distinctive leaf pattern . The map shows the different species that developed in the wild.

when plant breeders selected for color and size; the large-flowered, scentless plant had become the norm by 1880, and it is only in recent years that breeders have attempted to regain some of the lost charm of the wild plant. Breeding programs now often aim for smaller, scented cultivars, while retaining the bright colors. A recent breakthrough is the development of yellow-flowered cyclamen, achieved through a long process of selection.

Although *C. persicum* is tender, there are many hardy species, some of which will even withstand the harsh winters of the interior of the northern U.S. and Canada. Most familiar is the ivy-leaved *C. hederifolium*, a native of Italy and the Mediterranean eastward to Turkey, which has been enjoyed in cultivation from the 1500s. Its freely produced flowers appear in fall. The leaves appear after the flowers and are infinitely variable; enthusiasts collect those with the most unusual markings. They are normally marbled green and silver, but strains are now available with pure silver leaves. An old English name for this hardiest of cylamen was "sowbread," referring to the use of its tubers as pig fodder. John Gerard, in his Herball of 1597, recorded that a pregnant woman would miscarry if she were to step over the plants, and stated that his own plants were surrounded by a fence as a precaution against such tragedies!

The Flower of the Twelve Gods
The Ancient Greeks applied the name Dodecatheon to an unknown plant (possibly *Primula veris*), but the Swedish naturalist Carolus Linnaeus thought that it was an appropriate

name for a uniquely North American genus of multiflowered plants, whose flowers he regarded as "so many little divinities." It is commonly known as "shooting stars" and is found in damp places across much of the continent. The individual flowers resemble those of a cyclamen, with reflexed pink petals, but are borne in umbels, like the flowers of a polyanthus, on a

AMERICAN COWSLIP
Dodecatheon pulchellum
'Red Wine' (below) was
bred from a native
American species known
familiarly as the American
cowslip.

NATURAL SELECTION
Both in the wild and
naturalized in gardens,
Cyclamen hederifolium
(left) provides beauty and
interest from leaf or flower
throughout the year.

stem arising from a basal rosette of leaves. The eastern species *D. meadia* was first sent to Europe by John Bannister (d. 1692), an English clergyman who had been despatched to the Americas as a missionary by the gardening Bishop of London, Henry Compton. But the plants did not survive long at the bishop's residence, Fulham Palace. It was reintroduced in about 1740 by John Bartram (1669-1777), the first of the great American plant hunters, who sent specimens to his correspondent in England, Peter Collinson (1694-1768). It became just one of the many plants which were introduced by this partnership over three decades. *D. meadia* is easily grown in moist soil, as long as it is not stagnant.

*F*ROM CUPPED FLOWERS OF
MEADOWLAND AND WATER TO DOWN-
TO-EARTH DELPHINIUMS AND RAMBLING CLEMATIS.

BUTTERCUPS AND WATERLILIES
(RANUNCULACEAE, NYMPHAEACEAE)

Clematis florida 'Sieboldii'.

*F*ields, meadows, and grassy places teeming with golden buttercups are found throughout the northern temperate regions of the world. But few of the 400-plus species of *Ranunculus*, the main genus of the family Ranunculaceae, have become popular garden plants. They are, in the main, either surprisingly difficult to cultivate, or too "weedy". However the family also includes anemones, delphiniums and clematis, which are far more successful in cultivation. The genus name *Ranunculus* comes from the Latin, *rana*, meaning "frog". It is very

Delphinium cardinale *and another Californian species D. nudicaule, confound those who think all delphiniums are blue; a few other species are yellow. Originating in the chaparral of California, D. cardinale is easily grown in dry, sunny places elsewhere, and may reach 6 feet (2m) in height. It contributed its coloration to the red delphinium hybrids.*

Nymphaea 'Marliacea Chromatella' *is one of the many hybrid waterlilies raised in 1877 by the breeder Joseph Latour-Marliac at his nursery near Bordeaux, France. One of the parents was probably the yellow-flowered Nymphaea mexicana. Its canary-yellow flowers appear throughout summer, while its olive green leaves are handsomely marked with bronze. It does well in the shallow water of most garden pools; others need deeper water to flourish.*

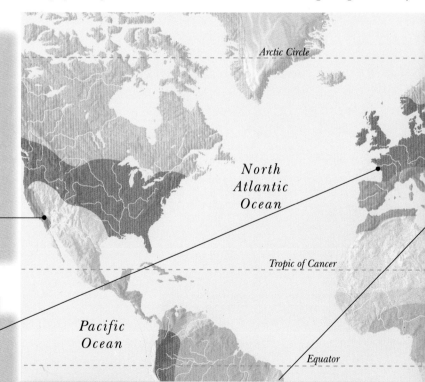

Arctic Circle

North Atlantic Ocean

Tropic of Cancer

Pacific Ocean

Equator

Anemone coronaria, *a simple wild flower from the Mediterranean basin, may have been the Biblical "lilies of the field" that rivaled the glory of King Solomon. It grows from a small tuber to flower in late winter and early spring, and lies dormant through the hot Mediterranean summer. Larger cultivars are grown commercially in northern Europe as cut flowers, while double forms were in cultivation before 1600.*

suitable for most European species, which are known as buttercups and crowfoots, since they are either aquatics or grow in damp places. The most widely cultivated species, *Ranunculus asiaticus*, however, occurs wild in dry places in the eastern Mediterranean. The single wild forms may be white, yellow, pink, or bright red, with a conspicuous mass of black stamens in the center. Double forms, known as "turban buttercups", were highly prized in the Turkish-based Ottoman Empire, mainly between the 1500s and 1800s. Some turban forms were bicolored or picotees. Turban buttercups grow from many-pointed tubers that become dormant in summer. They flower in late spring, or may be raised from seed to flower later in the year. The modern 'Bloomingdale' hybrids, with double flowers up to 3 inches (8cm) across, in many bright colors, reach only 8 inches (20cm) in height and can be grown as pot plants or for summer bedding.

The garland anemone

Anemone is a large genus in the buttercup family, with 120 species throughout northern temperate regions and mountainous tropical areas. The generic name is often supposed to come from the Greek *anemos*, meaning "wind," especially as some species are known as "windflowers". But it is more likely derived from the Semitic name for the demigod Adonis, *Naaman*. His spilt blood was said to have turned

Clematis rehderiana *is one of many small-flowered species of* Clematis *from Europe and Asia that are less frequently seen than the large hybrids. A vigorous climber,*

C. rehderiana *is best allowed to scramble through a tree, but it should always be planted where it is possible to smell the fragrant, cream-colored flowers which appear in large numbers from midsummer onward. The flowers are followed by attractive fluffy seed heads.*

THE FAMILIES' RANGE AROUND THE WORLD

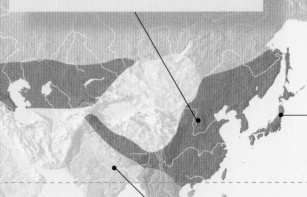

Indian Ocean

Anemone hupehensis *and its hybrids, often known collectively as* Anemone japonica, *have a long history of cultivation in both China and Japan,* *where they were often planted in graveyards. The plant was introduced to Europe in the 1800s and is appreciated for its late summer and fall flowers which may be pure white, pink, or red, single or double. In good soil, the plants soon spread to form large colonies, but also tolerate neglect.*

Nelumbo nucifera, *the sacred lotus of Asia, is valued in warm countries for its large, soft pink flowers, held above the water on long,* *stiff stems. The leaves are peltate, with a round blade supported by a central stalk; a glistening bead of dew or rain often sits in this natural cup—"the jewel in the heart of the lotus" that symbolizes the birth of Buddha.*

Key

 Tundra

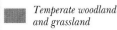 *Boreal coniferous forest*

Temperate woodland and grassland

 Mediterranean

Desert

Tropical woodland and grassland

Tropical rain forest

into one of the best known species, *Anemone coronaria*. Like turban buttercups, this is grown from a knobbly tuber, and is found wild in many parts of the Mediterranean, with bluish-purple, pink, or scarlet flowers. It may have been the brightly colored "lilies of the field" referred to by Jesus Christ in The Bible's *Gospel of St. Matthew*. The species name *coronaria*, meaning "of garlands," suggests that it was used as a garland flower in Classical times.

Desirable cultivated forms of this anemone began to spread through Europe in the 1600s. Allegedly a mean gardener in Holland, having raised an extremely fine strain, refused to let anyone else have plants. The Burgomaster of the town, who was among the many people who wished to grow them, cunningly arranged to pay a visit when the seeds were ripe. The Burgomaster arrived in his long, fur-trimmed civic robes, which gathered the fluffy anemone seeds as he swished past the beds.

Florists' favorites

Many anemones grown for cut flowers today are known as "De Caen anemones" from the town in northern France where they were developed. Their double counterparts, St. Brigid anemones, originated in Ireland. The named cultivars of these groups, such as the scarlet 'His Excellency', blue 'Mr Fokker' and the St. Brigid semidouble white 'Mt. Everest', must be grown from vegetatively propagated tubers. However, cultivars of mixed colors raised from seed are now available for the amateur grower.

The white-flowered northern European *A. nemorosa*, a woodland plant of quiet charm, has yielded several cultivars with double, pink or pale blue flowers. The blue 'Robinsoniana' was spotted by English gardener William Robinson (1838-1935), in a corner of the Oxford Botanic Garden during the late 1800s; its true origin has never been traced.

Eastern vigor

Most European anemones flower in spring, but the Japanese anemones from eastern Asia enliven borders in the late summer and autumn. The first to reach Europe were sent to the Royal Horticultural Society in England in 1844, by their collector Robert Fortune (1813-1880). He had found the flower growing on graves near the ramparts of the old city of Shanghai, and described it "is a most appropriate ornament to the last resting places of the dead." The flowers had many narrow, dull purplish-pink petals,

and were later recognized as a cultivar of the wild *A. hupehensis*. The wild form has single pink flowers and was not introduced to Europe until 1908. Fortune's plant was crossed with the white-flowered Himalayan species *A. vitifolia*, which had been introduced to Europe in 1829. The result is the many hybrids known today as *A.* x *hybrida*. They are vigorous, sometimes reaching 5 feet (1.5m), with pink or white flowers. The well-known cultivar 'Honorine Jobert' has pure white flowers with yellow stamens, and arose in 1851 as a mutation from a pink plant in the nursery of Monsieur Jobert at Verdun-sur-Meuse, France.

FLOWER OF THE WIND
Anemone nemorosa is the wild wood anemone of northern Europe, where it often grows in large drifts. The flowers nod in a slight breeze, well earning it the name "windflower."

DELPHINIUM: THE LEAPING DOLPHIN

The tall and showy delphiniums, *Delphinium*, members of the buttercup family, were slow starters in the world of garden cultivation. Annual and biennial species were grown in the Middle Ages in Europe. But the first of the robust perennials so well known today did not reach western Europe until the late 1500s. This was *Delphinium elatum*, from the rough meadows of eastern Europe and Siberia. In the wild it grows in dense colonies that turn the landscape the classic delphinium blue, with small flowers in narrow spikes that may be 6 feet (1.8m) tall.

Delphinium means "little dolphin," from the curved spur on the flowers of wild species, which resembles a leaping dolphin. However the spur is largely lost in modern hybrids. As with most members of the Ranunculaceae family, delphiniums are poisonous. The biennial species *D. staphisagria* was grown in the Middle Ages for its seeds, which were ground into a poultice for head lice. The related wolfsbane or monkshood, genus *Aconitum*, could poison larger animals. *Cimicifuga*, the bugbane, was said to drive away bedbugs.

Delayed development

Double varieties of early delphiniums were known from the 1600s, but little was done deliberately to improve the species for cultivation for some three centuries. Wild species such as *D. formosum* and *D. grandiflorum* were introduced, hybridization occurred, but few gardeners took note. Then in the mid

DOUBLE SCORE
The single-flowered, wild forms of Ranunculus asiaticus (left) were transformed to perfect doubles by gardeners, first in Turkey and later in Europe. All grow from a claw-like tuber.

1800s, French nurseryman Victor Lemoine (1823-1911) and his English counterpart James Kelway began breeding programs. The most important hybrid from Lemoine was a semidouble with flowers 3 inches (8cm) across, named 'Statuaire Rudé'. Kelway's finest was the plum-colored 'King of Delphiniums,' which was much used in later breeding work.

By the early 1900s, fine delphinium hybrids were being created by the English nursery of Blackmore and Langdon, which still specializes in them today. Breeders concentrated on fine flower form, robustness, and hardy habit to produce superb specimens for the herbaceous border. 'Fanfare', introduced in 1960, has silvery mauve flowers 3 inches (8cm) across, on tall spikes. Another fine race, the Pacific Giants, was raised in California by Frank Reinelt, a Czech emigrant who moved there in 1925. Pacific Giants perform well in a soft climate like that of their homeland, flowering in the first year from seed, and often blooming twice or more each year when older. But they are shorter-lived in cooler climates. Their large flowers are available in many colors, including pastel shades, as in the dusky pink 'Astolat'.

Delphiniums are basically blue. But various cultivated strains contain shades from white to deep purple, and wild species have red or yellow flowers. The red species *D. cardinale* and *D. nudicaule* are natives of California, and breeders began to try to introduce their coloration into the large-flowered border hybrids.

An early success occurred in the 1920s, when a cross between a border hybrid and *D. nudicaule* produced the dainty 'Pink Sensation' with good, clear pink flowers. In 1953, Dr. R. A. Legro began the search for red delphiniums at Wageningen University, in the Netherlands. After many years, red seedlings appeared. In 1980 the program was moved to the Royal Horticultural Society's gardens at Wisley, in Surrey, England. Eventually it produced red and pink cultivars with the same stature and quality as the familiar blues. Unfortunately, the red cultivars have also inherited short-living features from their ancestors, and so are unsuitable for commercial supply.

THE JAPANESE GARDEN

Dry courtyard garden, Japan.

The Japanese philosophy of Shinto teaches that all natural things are equal, and encourages their veneration; particularly beautiful objects are revered as the homes of spirits. Aesthetics are all-important in the selection of the constituents of a garden—a rock is selected and placed with care and thought; the note of splashing water must be just right. When Buddhism arrived from China in 595 A.D. its principles were seamlessly incorporated to form a united horticultural tradition.

Most important of the new Chinese ideas was the concept of the landscape garden, in which garden features represent a greater, often symbolic landscape. Rockwork became more elaborate and water was widely used, although it

was sometimes represented by a dry "watercourse" of carefully selected pebbles. Plants were chosen for form and texture, although a few favored flowers, such as tree peonies and chrysanthemums, might be grown in pots and brought into the garden when in flower. Moss, ferns, and grasses were the preferred low-growing plants, while evergreens provided stability and structure. Flowering trees and shrubs provided seasonal color, as did fall foliage, especially from the delicate Japanese maples (*Acer palmatum*). The main dwelling, as well as the bridges, pavilions, and tea houses, was considered to be an integral element of the garden, indicating an inhabited landscape.

The perfect place for tea

Contemplative harmony is the underlying concept of Zen gardens, which first appeared in the 15th century. These are often courtyard gardens where intricate patterns are raked in sand or pebbles around a few rocks, providing endless opportunity for the mind to ponder their meaning while contemplating the infinite. One of the most famous, Ryoan-ji, near Kyoto, is called the "Garden of Crossing Tiger-Cubs." It consists of five groups of stones, surrounded by a little moss, set amid raked quartz sand. The contemplative mind can read them as islands in a quiet sea, or the iceberg tips of the conscious emerging from the subconscious.

The tea ceremony is intended to induce contemplation and harmony, and it requires the provision of special garden features: a tea house approached by stepping stones to avoid damaging the moss carpeting the ground, a lantern to light the way, and a water-filled basin

HARMONY
The large-scale garden in Kyoto (above) recreates an ideal landscape with rockwork and ponds to suggest lakes and mountains.

CHEAP ALTERNATIVE
Saiho-ji (moss garden) in Kyoto (left) dates from the 1800s when the local monastery could not afford other plants.

for washing the hands. Developed in the 16th century, the tea ceremony demanded sobriety in all things, including the garden. Plants, which were mostly evergreens, were rigidly trained and clipped to provide a background that changed as little as possible with the seasons.

Many later gardens, such as those at Katsura Imperial Villa in Kyoto, incorporate features from all these traditions within a landscape.

RYOAN-JI, KYOTO *(right) A dry landscape without water, trees, or shrubs, but with 15 rocks in groups of five, three, and two, is nature in the abstract, a place of utter simplicity for profound contemplation.*

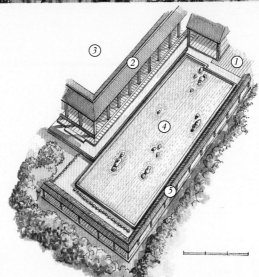

1. *Entrance*
2. *Covered veranda from which to view the garden*
3. *Guest house with Buddhist altar*
4. *Raked white sand*
5. *Screening wall 5 feet (1.5m) high*

CLIMBING CLEMATIS

Some 230 species of *Clematis*, another genus of the buttercup family, occur throughout North America, Europe and Asia, with a few species on African and New Zealand mountains. The name is from the Greek *clema*, "tendril," and was used in ancient times for several climbing plants. The clematis' climbing habit and showy flowers made it popular from the late 1500s, especially in the "English cottage garden" style.

The reddish-purple *Clematis viticella* from southern Europe was one of the first to enter cultivation. It is still grown today, and its small flowers suit less formal situations than the large-flowered hybrids descended from it.

Clematis cirrhosa is another Mediterranean species, cultivated in northern Europe by the late 1500s. Its creamy bells defy the worst winter weather by flowering from December to March. Cultivars include var. *balearica* from the Balearic Islands, with dissected foliage, and the recent 'Speckles' with heavy dark-red patches. All are hardy, down to about 5°F (minus 15°C), but prefer a sheltered wall.

Eastern input

Large, flat-flowered clematis species from China and Japan arrived in Europe from the late 1700s, beginning a new phase of development. The first to arrive, from Japan in 1776, was *C. florida* 'Alba Plena'. It was an old garden cultivar, rather than the wild species which occurs in western China. It has double flowers with many greenish-white petals, and would become a parent of many double cultivars, but is now seldom grown. Another striking arrival in the early 1800s was *C. florida* 'Sieboldii', still widely

BORDER HEIGHT
The blue spires of delphiniums (right) are an essential feature of traditional herbaceous borders. The black center of some cultivars is known to growers as the "bee".

grown for its bicolored flowers. The anthers have enlarged into purple, petal-like parts that lie flat against the white outer true petals. But it needs a favored site, since it lacks the vigor of many clematis hybrids.

Clematis florida and *C. patens*—another species from Japanese gardens—bear flowers on stems that are in their second season of growth. This trait has been passed to many of their hybrids. In contrast, the Chinese species *C. lanuginosa*, introduced to Europe in 1850 by Robert Fortune, from scrubby hillsides near the city of Ningpo, bears its starry, azure-blue flowers on first-season shoots. This feature of whether a clematis flowers on its older or newer growth is important for appropriate pruning!

Jackman's purple

Clematis x *jackmannii*, was produced in 1868 at Jackman's Nursery in Surrey, England, by crossing *C. lanuginosa* and the dark European cultivar *C. viticella* 'Atrorubens'. The velvety, deep purple-red flowers first appeared in 1862, and the plant caused a sensation when exhibited in 1863. It is vigorous, long-lived, and flowers prolifically on new growth in late summer. It soon became the standard against which newer cultivars were measured.

This success spurred a wave of clematis hybridization across Europe, which displaced many of the earlier cultivars. A few survive, such as 'Nellie Moser', its pale pink flowers striped with darker central marks, and 'Perle d'Azur', with large, pale blue flowers.

In the late 1800s, hybrid clematis with pendulous, more bell-shaped flowers were

PURPLE POPULARITY
Clematis 'Jackmannii' (below) has been popular from its introduction in the 1860s.

developed from the American species *C. texensis*. This had been introduced to Europe in 1868 and championed by Max Leichtlin (1831-1910) of Baden-Baden, Germany, but it soon spread farther afield, to Eastern Europe and back to North America.

The cultivar 'Etoile Rose' came from the French nursery of Lemoine. The deep velvety-red 1975 cultivar 'Niobe' was raised by the breeder Wladislaw Noll in Poland.

HARDY WATERLILIES

Botanists consider that waterlilies, family Nymphaeaceae, along with magnolias, are among the most primitive flowering plants. They may have evolved more than 100 million years ago, when dinosaurs roamed the land. The name is derived from the Greek *nympha*, meaning "water-being". In more recent history, waterlilies have been admired and even venerated for at least 6,000 years, inspiring

WATERGARDEN ART
The hybrid water lilies produced by Marliac, such as 'Escarboucle' (above) were planted by the artist Claude Monet in his garden at Giverny, France, and inspired a series of paintings.

PRIMITIVE ATTRACTION
The large flowers of water lilies are adapted to pollination by beetles. After pollination the fruit sinks and matures under water.

works of art from Ancient Egyptian tomb paintings to the imagery of French impressionist Claude Monet (1840-1926). Monet's waterlilies, in his garden at Giverny, northern France, were hybrids and selections of hardy species of *Nymphaea* from northern Europe and America. Most would have been raised by his compatriot, Joseph Bory Latour-Marliac (d.1911), at his nursery at Le Temple-sur-Lot, western France. Marliac was first to hybridize *Nymphaea* species, from 1879. He used several forms—the white-flowered, wild European *Nymphaea alba*, found even in the coldest areas of Scandinavia and therefore fully hardy; its var. *rosea* with pink flowers from Sweden; the fragrant American *N. odorata*; the dwarf *N. tetragona*; and the yellow *N. flava*.

Marliac set about the intricate task of pollinating the floating flowers and then collecting the sunken seed capsules for sowing. Over three decades, he developed the colored forms now seen in water gardens throughout northern temperate regions. One of his most outstanding cultivars is the bright red 'Escarboucle', released to commerce in 1909. It thrives in water less than 2 feet (60cm) deep and so suits small garden ponds.

Marliac claimed that the redness of 'Escarboucle' came from the tender Indian species *N. rubra*. But it is now thought to be the result of careful selection from *N. alba* var. *rosea*—and that Marliac was trying to deceive

AFRICAN BLUE
Nymphaea caerulea, seen here in a painting by Redouté (see page 179), is a blue water lily found throughout Africa; other species occur throughout the tropics.

would-be competitors. Also, Marliac never revealed his hybridization techniques, and it was believed that his secret died with him. Recently, however, it was discovered that he only released sterile cultivars, from which breeding was impossible, in order for his nursery to be the sole supplier.

Marliac's waterlilies continue to dominate the aquatic gardens of Europe. But equally fine hybrids and cultivars, often with striking colors, are now being developed in North America. A foremost breeder is Dr. Robert Kirk Strawn of Texas, U.S. His breeding program avoids Marliac's sterile cultivars and focuses on original species and fertile cultivars. Among his successes are the yellow-flowered 'Charlene Strawn', and the bright pink, multi-petaled 'Nigel'.

Tender waterlilies

To the Ancient Egyptians, the blue Nile waterlily, *N. caerulea*, was almost holy. From the Fourth Dynasty, some 6,000 years ago, it was widely depicted on temples and tombs, and on household items such as furniture and pottery. Scenes show slaves presenting the flowers to guests, to hold or place on their heads—possibly a symbol of peaceful intent. The flowers were also used as garlands on mummies of the rich and famous. Perfectly preserved flowers of *N. caerulea*, and the white *N. lotus*, were found in the sarcophagus of Pharaoh Ramses II, who died more than 1,500 years ago.

Growers have never succeeded in transferring this blue coloration of tropical waterlilies to the hardier cultivars of more northern regions. It remains one of the great challenges of plant-breeding. Later the name "lotus" was applied to the Asian plant *Nelumbo nucifera*. This was introduced to the Nile delta from India in about 525 B.C., when the Persians invaded Egypt, and the name was transferred. Most parts of this sacred flower are edible: the leaves make green vegetables; the starchy, energy-rich rhizomes are pickled for winter consumption; and the seeds are a delicacy, raw or cooked. The seeds are also extremely long-lived in mud or swampy peat, germinating successfully even after many centuries.

Nelumbo nucifera (family Nelumbonaceae) grows as a long creeping rhizome at the bottom of a pond or river. Strong stems emerge, bearing rounded leaves attached at their centers to the stalks. The globular flower is held well above the surface, with large, loose petals that are usually bright rose-pink, although cultivars

have paler, darker and double flowers. The lotus is considered sacred in Asia because Buddha was born from the heart of its flower. Hence the mantra whose repetition earns merit for the believer: *Om Mani Padme Hum*, "Hail to the Jewel in the Heart of the Lotus."

Giant of the Amazon

The backwaters and tributaries of the Amazon basin are home to the giant waterlily, *Victoria amazonica*. It was described for science in 1802, but was only named in 1832, by German botanist Eduard Poeppig (1798-1868). He chose *Euryale amazonica*, believing it to be a close relation of the prickly waterlily from India, *E. ferox*. In 1837, British explorer Sir Robert Schomburgk (1804-1865) saw the huge plants growing in Guyana. He believed that he had found a new species, and named it after his Queen: *Victoria regia*. However the hasty patriot soon realized that the species already had a formal scientific name. In yet another twist to the tale, further study showed that it was not quite a member of the genus *Euryale*, so the generic name *Victoria* could stand. But the first species name, *amazonica*, had to be reinstated. Finally the correct botanical name was established as *V. amazonica*.

The flower of this vast plant is composed of many whorls of petals, and may be more than 1 foot (30cm) across. On the first day it opens white, releases strong perfume and closes; on the second it re-opens as flushed pink but without the scent; finally it fades and sinks beneath the water. The scent attracts beetles as pollinators, which the flower closes to trap overnight, before releasing them on the second morning. Perhaps even more remarkable than the flowers are the huge rounded leaves, more than 6 feet (1.8m) across, with up-turned edges. They float on and are supported by their criss-cross structure of thick veins. In 1849 Joseph Paxton, the Duke of Devonshire's intensely practical head gardener, built a special conservatory for these giant waterlilies, which had arrived as seeds just three years before. He undertook experiments to test the strength of the leaves, and found that they could support his eight-year-old daughter Annie with ease.

Victoria amazonica is a popular feature of large greenhouses and botanic gardens around the world. It is a true perennial, but usually grown as an annual and reaches flowering size within a few months. It requires at least 3 feet (1m) of water and a temperature of 80°F (26°C).

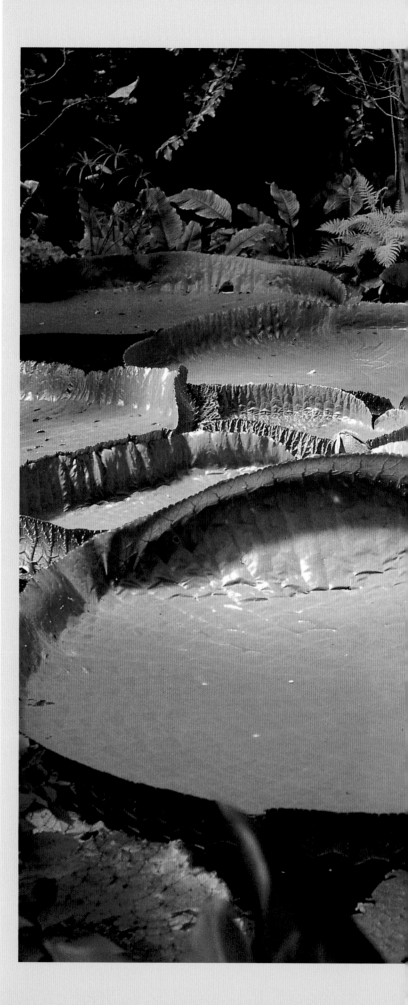

From waterlily to glasshouse

Sir Joseph Paxton (1803-1865) was an endlessly active character, full of energy and insight, who began as a humble gardener and became a British Member of Parliament. For a time he was the Duke of Devonshire's head gardener at Chatsworth House, when he also designed public parks and wrote books on dahlia cultivation.

Paxton was already noted for his skill in greenhouse design when he was asked to plan a building that could house Britain's Great Exhibition of 1851, in Hyde Park, London. The event intended to show off the wealth and products of the British Empire. At the time, Paxton was growing the giant waterlily, *Victoria amazonica*, in a tank and conservatory that he had designed himself at Chatsworth. He realized how the longitudinal and transverse veins in the huge leaf could be adapted to glasshouse construction, to achieve the same effect—rigidity and strength coupled with a graceful pattern effect.

Paxton was sitting in a board meeting at a railway company one day, when he was suddenly inspired to sketch his first design on a sheet of blotting paper, and completed the detailed plans for the Crystal Palace, one of the world's largest buildings, within ten days, just in time for the event.

Greenhouses had been used to protect tender plants since the late 1500s. But most were dingy, dank places where plants survived rather than flourished. The Crystal Palace revolutionized their design. Glass-walled conservatories became a feature of the late 1800s throughout Europe and North America. This was assisted by better glass-making techniques, which allowed larger, clearer panes and more light within. Similar structures, but on a more mundane level, soon became popular for commercial crops of flowers or fruit. The Crystal Palace itself was moved from Hyde Park to Sydenham, South London in 1854, but was destroyed by fire in 1936.

"...No man knows through what wild centuries roves back the rose."—Walter de la Mare. The evolution of the modern garden roses started in the eastern Mediterranean or the Near East nearly 4,000 years ago.

THE ROSE FAMILY
(ROSACEAE)

Rosa 'Peace'

The earliest known depictions of roses are in frescoes dating from around 1500 B.C., in the Minoan palace of Knossos on Crete. Their identity is not certain, but it is probably a hybrid derived from the wild species *Rosa gallica* and *R. phoenicia*, which both occur in the eastern Mediterranean. Later crosses between these parents gave rise to the Damask roses (*R. x damascena*), which were brought to northern Europe by Crusaders from the gardens of Damascus. Roses were highly valued by the Saracens, and Persian poetry has many references to them. A damask rose was planted on the grave of the poet Omar Khayyam (c. 1050-1123). His long poem, the *Rubaiyat*, translated into English by Edward Fitzgerald in

RAMBLING WILD

Arctic Circle

North
Atlantic
Ocean

Pacific
Ocean

Rosa rugosa, *the "wrinkled rose," is commonly naturalized on the shores of eastern Canada and the U.S., as well as around the Great Lakes. It is also found wild in open places across much of the eastern Asian coast from Korea and Japan to Kamschatka. Large single flowers are followed by tomato-like red hips, and the foliage may become bright yellow in fall. A pure white cultivar, 'Alba', is very beautiful, while the semidouble 'Roseraie de l'Hay' is particularly strongly scented.*

Rosa canina, *the "dog rose" is the common wild species of woodlands and hedgerows in northern Europe. Like all wild roses the flowers are followed by hips rich in Vitamin C which can be made into jelly. Although a plant in the grounds of Hildesheim Cathedral, Germany, is reputed to be over 1,000 years old, it is seldom grown in gardens except as a rootstock for budding hybrid cultivars or other rose species.*

the 1800s, contains many allusions to the fleeting beauty of roses:

"Each Morn a thousand Roses brings, you say;
Yes, but where leaves the Rose of Yesterday?"

A seedling from the plant on his grave in Nishapur, Iran, was planted on Fitzgerald's grave in England, and is known as 'Omar Khayyam.' Pleasantly fragrant, damask roses are the source of essential rose oil, which is distilled from the young buds; roses are grown in large quantities for this purpose in Bulgaria.

By the 15th century roses were established features in medieval gardens. When civil war broke out in England in 1455, the rival factions chose roses as their emblems: white for the House of York and red for the House of Lancaster. The Wars of the Roses ended with the victory of Henry Tudor (King Henry VII) over Richard III in 1485. The ancient rivalry continues in the annual "Roses" cricket match played by the county teams of Yorkshire and Lancashire.

The red rose of Lancaster is believed to be *Rosa gallica* 'Officinalis', a bright red, semi-double cultivar of great antiquity. Certainly grown by the ancient Greeks and Romans, it was probably also cultivated in Ancient Persia. At some time in the Middle Ages, *R. gallica* 'Officinalis' must have produced a mutant shoot of pale pink flowers, streaked and spotted with red. This plant, *R. gallica* 'Versicolor', became known as *Rosa Mundi* ("the rose of the world") and remains a widely grown shrub.

White roses, *R. x alba*, were also grown in classical times and are probably a hybrid between *R. x damascena* and a white-flowered dog rose, *R. canina*. A semidouble cultivar, 'Semi-plena', is the white rose of York. The fully double 'Maxima' is the Jacobite rose, whose flowers became the emblem of supporters of

Key

Tundra

Boreal coniferous forest

Temperate woodland and grassland

Mediterranean

Desert

Tropical woodland and grassland

Tropical rain forest

Rosa foetida *is a native of western and central Asia, where it was also cultivated. Yellow roses are, without exception, native to Asia, so its introduction to Europe in the 1500s (via Austria—it has been called the Austrian briar ever since) was sensational. 'Bicolor', shown here, is an orange cultivar also known as 'Austrian Copper'. About 1900 it gave its bright colors to modern roses, but also contributed its susceptibility to the fungal disease blackspot.*

Rosa moyesii *has perhaps the brightest red flowers of any rose, borne on strong arching stems that may reach 10 feet (over 3m) in height.* They are followed by a prolific crop of elongated, bristly hips. Its size and the ferocity of its thorns make it suitable for wilder parts of the garden. A native of western China, it was introduced to horticulture in 1903.

Tropic of Cancer

Indian Ocean

Equator

Rosa banksiae *is named after the wife of the great Sir Joseph Banks, who introduced it from China in the late 1700s. The first clones to reach Europe from Canton were cultivated double forms, including the familiar butter-yellow* 'Lutea' shown here. The wild species comes from western China, where it was discovered much later. A very vigorous climber, R. banksiae likes to scramble over a tree, or in cold climates, against a building for shelter.

CHINA ROSE
Rosa chinensis (left),
contributed its repeat-
flowering habit. Painting
by Alfred Parsons

Prince Charles Edward Stuart (1720-88) and his claim to the British throne. Bushes of it are planted around the lonely monument at Glenfinnan in Scotland that commemorates Bonnie Prince Charlie and the doomed 1745 uprising against English rule.

The cabbage rose, *R. x centifolia*, the moss roses (forms of *R. x centifolia* with mosslike hairs on the calyces), the Damasks, Gallicas, Albas, and their numerous cultivars form the so-called "old-fashioned" roses. Their rounded flowers, in shades of white, pink, red, and purple, are renowned for their perfume, but sadly only appear for a brief, but glorious period, in early to mid summer.

The first revolution

Some time—no precise records exist—before the 10th century A.D., in western China, a wild rose that flowered throughout the summer was brought into domestic cultivation. Cultivated roses soon became as popular in China as they were in Europe, with single or double cultivars in shades of pink, copper, crimson, and yellow being grown. These began to trickle into Europe from the early 1700s onward, and were given the name *R. chinensis*, although some are, in fact, hybrids with *R. x odorata*. Typically small, twiggy bushes, the plants are quite different to the

CLASSIC BEAUTY
The strongly scented
Damask rose 'Madame
Hardy' (left) was raised in
France in 1832.

The artist and the empress

Pierre-Joseph Redouté (1759-1840) lived in an age when wealthy patrons were prepared to pay artists to record the plants in their collections. Redouté worked for a series of great French botanists, illustrating their works with his fine engravings, which were then individually hand-colored. However, his most famous association was with the Empress Joséphine and her garden at Malmaison near Paris.

Joséphine had married Napoleon Bonaparte in 1795, and bought the dilapidated Malmaison estate in 1799. She had the gardens restored, and commissioned Redouté to record the plants, which were obtained from around the world, resulting in what is regarded as his finest work: *Jardin de la Malmaison* (1803-5). Despite the war between Britain and France, Joséphine imported plants from England, including many of her favorite roses. The beautiful pale pink Bourbon rose, 'Souvenir de la Malmaison', commemorates her garden, although it was not introduced there until 1843. The illustrations of old-fashioned roses in Redouté's three-volume *Les Roses* (1817-24) have been endlessly reproduced. An artist to the last, he suffered a fatal stroke when examining a lily he was about to paint.

HARKING BACK TO THE
RENAISSANCE, THE GARDENS AT
SISSINGHURST RELATED TO THE BUILDINGS, AND WERE FORMAL IN PLAN,
BUT THE PLANTING SCHEMES DISPLAYED A DELIGHTFUL INFORMALITY.

AN ENGLISH GARDEN

*Perspectives from the
Rose Garden*

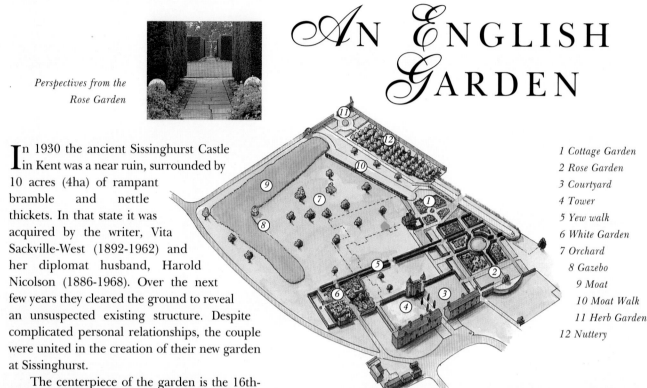

1 Cottage Garden
2 Rose Garden
3 Courtyard
4 Tower
5 Yew walk
6 White Garden
7 Orchard
8 Gazebo
9 Moat
10 Moat Walk
11 Herb Garden
12 Nuttery

In 1930 the ancient Sissinghurst Castle in Kent was a near ruin, surrounded by 10 acres (4ha) of rampant bramble and nettle thickets. In that state it was acquired by the writer, Vita Sackville-West (1892-1962) and her diplomat husband, Harold Nicolson (1886-1968). Over the next few years they cleared the ground to reveal an unsuspected existing structure. Despite complicated personal relationships, the couple were united in the creation of their new garden at Sissinghurst.

The centerpiece of the garden is the 16th-century brick tower, around which a series of separate garden "rooms" spread out. The garden's skeleton is strictly formal, and was laid out by Harold Nicolson; both he and Vita greatly admired the gardens of the Italian Renaissance. Within the spaces imposed by the garden's structure, Vita created sumptuously informal gardens, in which plants of all kinds were blended. She wrote that Sissinghurst combined "the strictest formality of design with the maximum informality in planting."

In the 1920s Harold Nicolson was posted to Tehran, resulting in the introduction of a Persian (Iranian) element in Vita's gardening, She assembled a fine collectionof old-fashioned roses, which had largely disappeared from English gardens, likening their muted colors to those of Oriental carpets. The roses were mixed with other plants, but always with a close eye to the color scheme—following the example set by the English gardener, Gertrude Jekyll. A plant was allowed to flop or self-sow, but if it looked out of place it was removed.

As a contributor to the gardening columns of *The Observer*, Vita passed on many of her techniques and thoughts to her readers, in a characteristic combination of practicality and poetry. In the winter of 1950 she told her readers of "the pale garden that I am now planting, under the first flakes of snow." The White Garden would contain only white flowers and gray foliage; it became the most famous of Sissinghurst's garden rooms, and has been imitated all over the world.

Today, details of the planting may have changed, but the garden is essentially much as Vita created it, a triumph of artistry and controlled informality.

FORMAL STRUCTURE
The overall scheme at Sissinghurst (above), incorporates a series of "garden rooms" centered on the Elizabethan brick tower (right). Each "room" is a self-contained unit with a particular function or theme.

A GARDEN IN WHITE
Vita planted the garden (left) in winter, "hoping that the gray ghostly barn owl will sweep silently across a pale garden, next summer in the twilight…"

Rosa x *odorata* (*R. chinensis* x *gigantea*) *is an ancient hybrid from China, which brought vigor and a tea-like fragrance to Europe.*

Hybrids between R. x *odorata and Hybrid Perpetuals, combining genes from Europe and Asia produced the repeat-flowering Hybrid Tea roses, such as 'Fragrant Cloud' (above).*

R. foetida *'Persiana' gave breeders the material to add brilliant yellow colors to hybrid roses.*

Rosa chinensis brought repeat flowering and high-centered flowers to garden roses.

Hybrid Perpetual roses such as 'Souvenir du Docteur Jamain', resulted from crosses of Bourbon roses with R. chinensis, *which improved their repeat-flowering qualities.*

Modern Floribundas such as 'Korresia' combine the clustered heads of R. multiflora *with the pointed shape of Hybrid Teas and the bright yellow of* R. foetida—*incorporating genes from at least five species of wild rose.*

Old-fashioned roses such as <u>Rosa gallica</u> 'Officinalis' are derived from species found wild in Europe and western Asia, and flower only once a year.

Bourbon roses such as 'Mme. Isaac Pereire' are the products of crosses between European roses and <u>Rosa chinensis</u>, and bloom at least twice a year.

<u>R. multiflora</u> from Japan has many small flowers clustered together, and was a parent of the first Floribunda roses.

ℰVOLUTION OF THE ℳODERN ℛOSE

The development of the modern rose has involved many species and the efforts of plant breeders for over 200 years. Many thousands of cultivars are the result, in every color except true blue.

vigorous European roses, and many of the original clones died out long ago.

The importation of four distinct clones of *R. chinensis* to Europe between 1789 and 1824 literally reshaped the rose. For the first time, breeders saw the possibilities of remontant (repeat) flowering roses. A small, pink-flowered cultivar of *R. chinensis*, known as 'Parson's Pink China', arrived in England in 1789 and was soon being grown throughout Europe. It reached North America in 1800, and on both continents gardeners began to raise seedlings from it and other china roses. The real breakthrough, however, was an accident.

In about 1810 'Parson's Pink China' was sent to the French colony on the Indian Ocean island of Réunion (then called Isle de Bourbon) and was planted as hedges by the settlers. Nearby was a bush of *R. x damascena* 'Bifera', by chance the only old European rose to produce a few blooms throughout the year. The two crossed naturally, producing the Bourbon rose which, flowering for the first time in 1817, combined the magnificence of the European roses with the repeat-flowering of the China roses. The floodgates of hybridization were opened. Bourbon roses rapidly became an important group, remaining widely grown today. Many bear the names of fashionable French ladies, now otherwise quite forgotten: 'Mme Pierre Oger' has a blush-pink flower, while 'Mme Isaac Pereire' combines a huge, magenta-rose flower with an incredibly rich scent.

The Hybrid Perpetual roses were developed by back-crossing with China roses, firmly establishing the remontant habit. Other repeat-flowering hybrid groups included the Noisettes,

FRENCH GLORY
'Gloire de Dijon' (left) was one of the first yellow hybrid roses, and rapidly achieved popularity; it is still widely grown.

INTERNATIONAL
SYMBOL
*'Peace' (above) displays
the classic Hybrid Tea
rose shape in its large,
pointed flowers, borne
singly on stout stems.*

and Hybrid Teas. 'Gloire de Dijon,' with its soft creamy-peach flowers, is a direct descendent from the Noisettes, first appearing in 1853.

The mid-19th century saw the development of the first tea roses. These were named for their scent, which suggested the fresh aroma of a chest of fine China tea. Tea roses were derived in part from *R. x odorata*, another Chinese species. The crosses introduced fine shades of yellow, but also proved to be so tender that most tea roses are no longer grown. Crossed with Hybrid Perpetuals, tea roses produced the Hybrid Teas, the first of which, 'La France,' was raised in 1867. Hybrid Teas are characterized by pointed buds and flowers held singly at the ends of stiff stems—in less than 50 years the Chinese influence had completely transformed the flat-flowered old European roses.

Paralleling this development of the "HTs" was the creation of another group, the Floribundas, in which the stems bear many smaller flowers. Originating as a hybrid between a China rose and *R. multiflora*, a Japanese species with small white flowers, the Floribundas at first had rounded flowers, but later crosses introduced the pointed Hybrid Tea shape.

The second revolution

By the 19th century, there was still no bright yellow hybrid rose, although yellow roses were readily available. *R. foetida*, for example, was a wild species with bright yellow flowers, which had been cultivated in Islamic gardens for centuries, and in Europe since the 1500s. Brought by the Dutch plantsmen Clusius from Vienna to Holland in 1583, it has been known as the Austrian Briar ever since. Another cultivar, *R. foetida* 'Bicolor', had orange-red petals shaded yellow externally, giving rise to its name of Austrian Copper. And a double form, 'Persiana,' had arrived from Iran in 1838. However, despite its promise, *R. foetida* proved to be reluctant to form hybrids.

Eventually, persistence brought success to Joseph Pernet-Ducher of Lyon. Between 1883 and 1888 he made thousands of crosses of *R. foetida* with Hybrid Perpetuals. In all that time, only one cross resulted in fertile seeds. From these two, seedlings germinated. One was sterile, but the other went on to contribute the sought-after yellow genes to its offspring. Ducher released the first yellow hybrid, 'Soleil d'Or,' in 1900. Crossed with Hybrid Teas and Floribundas, the vivid yellows and oranges derived from *R. foetida* became an integral part

the derivatives of a cross between *R. chinensis* and the white-flowered *R. moschata*, a climber from western Asia. The cross was made in 1802 by John Champneys of Charleston, South Carolina, from a plant of 'Parson's Pink China.' The plant had been obtained from a nursery owned by the Noisette brothers. The resulting 'Champneys' Pink Cluster' is a climber with large bunches of double, pink flowers, produced throughout the summer. It quickly became very popular in the United States. Philippe Noisette raised seedlings from this plant in Charleston and sent the best, the first Noisette rose, to his brother Louis in Paris. It became popular in Europe and many seedlings were raised from it. Some were crossed with another China rose, 'Parks' Yellow China,' bringing yellow to these and the other new classes of roses—the Teas

of the gene pool. Unfortunately, with the yellow coloration came a more sinister gene; roses derived from *R. foetida* are susceptible to the fungal disease blackspot, which causes disfiguring leaf loss and poor growth.

Crosses between wild species and garden hybrids have resulted in many other groups of roses, such as Ramblers, Sweet Briars, and Hybrid Musks. A relatively new group, known as English roses, appeared in the 1970s, and quickly become popular throughout the world. First raised by the English nurseryman David Austin, English roses combine the scent and shape of old-fashioned roses with the repeat flowering and colors of modern Hybrid Teas and Floribundas.

'Peace' in our time

The most widely planted rose of all time is the Hybrid Tea, 'Peace.' Its classically shaped flowers have petals of pale yellow edged with soft pink and are held singly above glossy, dark green foliage. The rose began as a seedling in the beds of the Meilland family nursery near Lyons, France. In 1935, after its first flowering, it was selected for further trial. By 1939 a larger stock had been built up. The bushes, under the breeder's code number '3-35-40', were in full flower when a party of international visitors arrived in June. Of all the Meillands' roses, '3-35-40' was the one they liked best, and several growers requested propagating material for that autumn. As Europe became embroiled in the chaos of the Second World War, the Meillands sent out parcels of bud wood to contacts in Germany and Italy. A third parcel left France on the last commercial flight to the U.S. All communication ceased. In France, the rose was released under the name 'Mme. A. Meilland' after the matriarch of the Meilland rose-growing dynasty. In Germany, it was named 'Gloria Dei' (Glory be to God) and in Italy, 'Gioia' (joy). In America, '3-35-40' flourished and April 29, 1945 was set as the day on which it was to be named by the American Rose Society. As white doves were released at the ceremony in Pasadena, California, a statement was read out: "We are persuaded that this greatest new rose of our time should be named for the world's greatest desire, PEACE." A few weeks later, the delegates to the first meeting of the United Nations in San Francisco found a flower of 'Peace' in each of their hotel rooms, bearing the message, "This is the rose 'Peace' which received its name on the day Berlin fell.

May it help to move all men of goodwill to strive for Peace on earth for all mankind."

FLOWERING CHERRIES

Although overshadowed by roses, many other members of the Rosaceae are important garden plants, often in a culinary as well as ornamental capacity. Apples (*Malus*), pears (*Pyrus*), plums, peaches, cherries (*Prunus*) and strawberries (*Fragaria*) all have the five-petaled flowers characteristic of the rose family. Members of all these genera find a welcome place in the flower garden, but none are more worthy than the flowering cherries of Japan.

Although many species of *Prunus* are indigenous to Japan, it is believed that the parent of most Japanese cherries, *P. serrulata* var. *spontanea*, the Yamazakura or wild cherry,

SRING RENEWAL
Flowering cherries, important in Japanese culture for centuries, are now a springtime feature throughout the temperate regions of the world.

was brought from China by Buddhist missionaries in the 6th and 7th centuries A.D. Groves were often planted around Buddhist shrines, and the white flowers were regarded as symbols of purity as well as of chivalry and knightly honor. Unlike many flowers reserved for the nobility, the cherry tree became a plant of the people. The Yamakazura has remained the most admired and revered of all, and is especially popular for cherry-blossom-viewing parties. These famous gatherings are recorded from the 1500s onward, when they were held by the emperor and nobility; today the cherry blossom season is an opportunity for anyone to picnic. Great parks were planted with cherry trees. At Koganei near Tokyo, for example, an avenue three miles (5km) long was planted in 1735. Some of the original trees survived into this century. From Japan the tradition of mass-planting has traveled around the world. One of the areas now famed for the beauty of its cherries is the Mall in Washington D.C.

Double-flowered cherry cultivars appeared about 1,000 years ago, and were usually planted as individual specimens in temple gardens, although their group name, *Sato-zakura*, means "village cherry." The oldest surviving cultivar is 'Fugenzo' ("goddess riding on a white elephant"); its unmistakable double, rose-pink flowers and coppery new foliage appear in 500 year old pictures. However, it did not appear in Europe until 1878, one of the first of many cultivars to be exported from Japan in the late 19th and early 20th centuries. Regrettably, its fine form was usurped by the vulgar 'Kanzan,' which became the most widely grown cultivar in Europe, apparently admired for its heavy crop of sugar-pink flowers. 'Amanogawa' ("milky way" or "celestial river") bears pale pink, double

flowers on stiffly erect branches, making it a useful tree for small gardens, while 'Ukon' bears flowers of a curious greenish cream from spreading branches.

POTENTILLA AND ALCHEMILLA

Most members of the rose family are woody, but a significant minority are herbaceous perennials. The genus *Potentilla* includes both woody and herbaceous species. Its English name, cinquefoil, was suggested by the five leaflets that make up the leaf of many species. Naturally occurring in mountains and uplands throughout Eurasia and North America, the shrubby cinquefoil, *Potentilla fruticosa*, is one of the most hardy of small shrubs. It is also extensively planted in other parts of the world, such as North America. Not surprisingly for a species with such a wide distribution, *P. fruticosa* is very variable in habit and color. Bright yellow, cream, and white variants are commonly found in Eastern Asia and, in 1920, Reginald Farrer (see p.41) collected seed of a red-flowered variant in upper Burma. It gave rise to orange-flowered plants, but it was not until the early 1970s that a red-flowered variety appeared in an English nursery. It was named 'Red Ace' and, despite being a weak grower, became the parent of many bright red, orange, and pink cultivars.

The name *Potentilla* means "little powerful one," from its medicinal uses, but although *Alchemilla* ("little alchemical one") also had medicinal properties, its name relates to more magical powers. To the medieval mind, dew was a magical substance, and so a plant that seemed to concentrate dew partook of that magic. The leaves of *Alchemilla* are rounded, their shallow lobes edged with little serrations. The leaves fold up at night and when they open in the morning dew drops sit upon the leaf surface and are often held on each of the marginal teeth. The many species are worth growing for this lovely feature alone, but in summer they produce minute lime-green flowers in fluffy masses, making an ideal foil for bright flowers grown around them. The most popular species is the large *A. mollis*, a native of the mountains of eastern Europe and western Asia, but smaller species, such as *A. conjuncta* from the European Alps, may better suit small areas. In many European languages, their common name ("Lady's Mantle" in English, *Frauenmantel* in German) is a Christian reaction to their reputed link with magic, as well as describing the shape of the leaf.

LADY'S MANTLE
The common name of "Lady's mantle" for Alchemilla species is inspired by the scalloped leaves (left). This species is A. volkensii from the mountains of East Africa.

TOUGH CHOICE
Potentilla fruticosa is an extremely hardy shrub with a wide distribution. Usually yellow-flowered, 'Tangerine' is one of several orange or red flowered cultivars raised from seed originally collected in Burma.

SHRUBS AND TREES FROM THE
ORIENT AND AMERICA, HANDSOME
OF LEAF AND EXOTIC OF BLOOM: MANY WERE TRANSPORTED FROM THEIR
HOMELANDS IN THE HANDS OF ENTERPRISING MERCHANT SEAMEN.

CAMELLIAS AND MAGNOLIAS

(THEACEAE AND MAGNOLIACEAE)

Magnolia grandiflora

The true origins of *Camellia sinensis* are blurred by its long history of cultivation in China, where its leaves have been dried and infused into tea by the Chinese for nearly 3,000 years. It probably came originally from the mountain region between China and India. The glossy, evergreen leaves and white flowers caught the attention of Europeans in the 1600s, and the Dutch soon created the first tea plantations in the Dutch East Indies (now Indonesia). Although the requirement of a warm, humid climate means that *C. sinensis* was never destined to be successful as an ornamental plant in Europe, it thrives in the southern U.S. where its elegant form is much admired.

The family name of Theaceae reflects the commercial value of *C. sinensis* as a tea crop, but the generic name *Camellia* comes from a Czech, Georg Kamel (1661-1706), who, when working in the Philippines as a Jesuit missionary,

Key

Distribution of Camellias

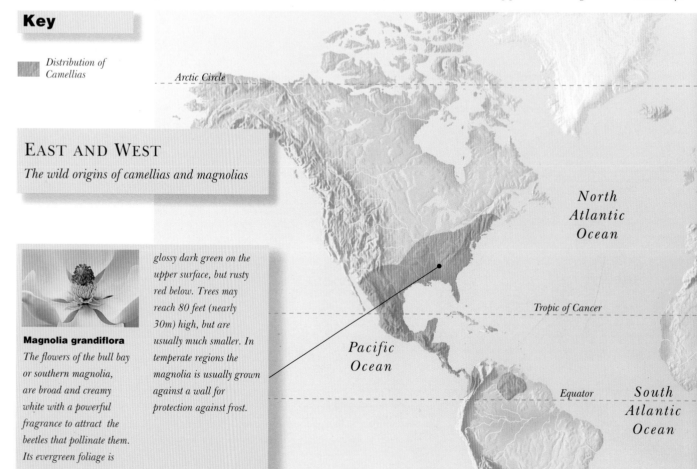

EAST AND WEST

The wild origins of camellias and magnolias

Magnolia grandiflora

The flowers of the bull bay or southern magnolia, are broad and creamy white with a powerful fragrance to attract the beetles that pollinate them. Its evergreen foliage is *glossy dark green on the upper surface, but rusty red below. Trees may reach 80 feet (nearly 30m) high, but are usually much smaller. In temperate regions the magnolia is usually grown against a wall for protection against frost.*

Arctic Circle

North Atlantic Ocean

Tropic of Cancer

Pacific Ocean

Equator

South Atlantic Ocean

collected some of the first specimens of Asian plants to reach Europe. His latinized name becomes *Camellus*, and, although seldom heard, the generic name should be pronounced "Camell-ia," not "Camee-llia."

Enterprising trade

China was very much forbidden territory to outsiders until the second half of the 1800s, when the Opium Wars between Britain and China of the 1840s forced it to open its doors more widely. Until then foreign trade was limited to a few ports ruled by special treaty agreements with European merchants, such as Canton (Guangzhou) and Macao. Few visitors went beyond the ports and their first impression of the indigenous flora was gained from the local gardens. Enterprising merchant captains foresaw a profit in selling such novelties back home and so the first Chinese plants to reach the West were cultivated varieties rather than wild species. The first *Camellia japonica* plants probably reached Europe via merchant seamen, and the plant's first flowering was in Essex, England during the winter of 1739.

European gardeners were ignorant of the natural growing conditions of these exotic new plants and placed the specimens in hothouses where they promptly died. In fact, in its wild form, *C. japonica* is a small tree from the mountain forests of Japan and Korea, and can withstand frosts to -28°F (-18°C). In north temperate areas it is still often grown under protection, as the flowers which open in early spring, are ruined by a light touch of frost and the year's display can be lost overnight.

There was no such problem in the southern part of North America, where the first *C. japonica* was introduced in 1785 at Middleton

Magnolia campbelli *is a forest tree from the lower slopes of the Himalayas in India, Nepal, Burma, and western China, bearing huge, pink or white flowers on leafless branches in early spring. Although the tree is relatively hardy, it needs careful siting as the flowers are destroyed by frost. Introduced to Europe in the late 1800s, it took up to 20 years for the first seedlings to flower.*

Magnolia denudata *The yulan or lily tree, is native to central China, but has long been valued in Japan for its pure white, deliciously fragrant flowers, which to Buddhists, symbolize purity. Although slow growing, it eventually becomes a medium-sized tree, which, like many other magnolias, prefers fertile, slightly acid soil that does not become too dry in summer.*

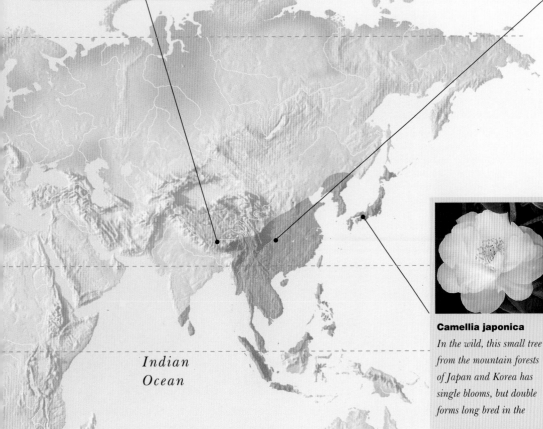

Indian Ocean

Camellia japonica
In the wild, this small tree from the mountain forests of Japan and Korea has single blooms, but double forms long bred in the Orient were introduced to the West in the late 1700s. The flowers can be ruined by frost, but *C. japonica* was later crossed with *Camellia saluensis* to produce the robust *C. x williamsii* hybrids.

SEAFARING BEAUTY
Camellia reticulata
'Captain Rawes' (above)
commemorates a British
sea captain who took the
first plants to Europe.

HARDY HYBRID
New cultivars are still
being developed from the
hardy hybrid Camellia x
williamsii (below),
introduced in the 1920s.

Place, South Carolina, by the great French gardener André Michaux.

The wild form of *C. japonica*, together with approximately 80 other species, has single flowers, but double forms in a range of colors from white to crimson have long been bred by the Chinese and Japanese. In 1792, John Slater, a director of the British East India Company, imported two double flowered forms to Britain, one white, the other streaked with red, and by the early 1800s these were established as favorite glasshouse plants throughout Europe and North America.

Packed with care

Hundreds of plants died on the long sea voyages from the East to Europe and America, until John Reeves, a British East India Company tea inspector based in Macao and Canton, revolutionized potting and packing methods. Reeves (1774-1856) became one of the most influential figures in the introduction of Chinese plants to Europe, not only because of the

number of his plants which arrived in good health, but also because he commissioned Chinese artists to make a valuable painted record of them. These paintings are now in the Royal Horticultural Society's Lindley Library in London and the British Museum.

Reeves selected the ships on which his delicate cargos would travel with care. He entrusted many of his plants to Richard Rawes, Captain of the East India Company vessel *Warren Hastings*, which carried to England the first *Primula sinensis*, *Wisteria sinensis*, and a beautiful, semidouble pink form of *Camellia reticulata*, which is now known as 'Captain Rawes'. *C. reticulata* is less hardy than *C. japonica* but it is more flamboyant. Its wild origins are in western China, but double forms have been cultivated for centuries in temple gardens. There are trees still living today, gnarled with great age, which continue to bear heavy crops of flowers. The 11th-century Chinese naturalist Chao-pi listed 72 distinct *C. reticulata* cultivars—the Chinese had perfected grafting

techniques—and gave the new cultivars fantastic names such as "Golden Heart Jewelry" and "Envious of Heaven's Height."

Exploring the hinterland

In a narrow 50 mile (80km) band separating China from Burma, are the valleys of four mighty rivers, Irrawaddy, Salween, Mekong, and Yangtze, and the steep mountain ranges that divide them. It is a tangled terrain, which provided rich hunting ground for many of the plant collectors of the early 1900s. One of these was a Scot, George Forrest (1873-1932), who, in the course of seven plant-hunting trips in western China between 1904 and his death in Yunnan Province, collected an enormous number of plants, many of which were new to science. Forrest managed to cover a wider territory than most, as he trained his Chinese assistants to take part in the collecting.

The turn of the century was a golden age for plant collectors as rich landowners, keen for a share in the new plants, would sponsor their trips. One of Forrest's backers was J.C. Williams (1861-1939) of Caerhays Castle, Cornwall, England, to whom he sent in 1917, seed of *Camellia saluenensis*, a species that had not previously been in cultivation. He had found the small shrub with 2-3 inch (5-7.5cm), white to red flowers widespread in Yunnan, and named it after the Salween River. The plant's fame, however, was sealed when Williams crossed it with *C. japonica*, and created a race of floriferous and very robust garden shrubs called *C.* x *williamsii*. The flowers of *williamsii* camellias are usually single or semidouble, and appear less formal than those of many *C. japonica* clones.

MAGNOLIAS: AN ANCIENT LINEAGE

While the camellia's origins are confined to Asia, the genus *Magnolia*, with 80 or so species, is represented in both Asia and Eastern North America. This distribution is shared with many other genera of woody plants extant in the Tertiary period when flowering plants first made a significant impact on the evolution of

Extent of ice caps in glacial period

Tundra

Temperate forest/ magnolia distribution

Desert and dry grassland

N-S mountain ranges

Transverse mountain ranges

Ancient survival

During the Tertiary period (65-2 million years ago) flowering plants, or angiosperms, became more abundant than the flowerless gymnosperms, and developed large forests in the north temperate region. Many of these early plants, including *Magnolia* and its close relation the tulip tree (*Liriodendron*), both of which are now found in Asia and North America, became widely distributed.

Major climatic change occurred toward the end of the Tertiary, and the world entered a glacial period, in which polar ice caps expanded and the temperate forests died. Primitive plants were eradicated from most of Europe and northern Asia, but in parts of China and North America, conditions remained warm enough for relict populations to survive. This split distribution can be explained by the presence in Europe and much of Asia of transverse mountain ranges, such as the Alps and Himalaya, whose subsidiary icecaps prevented southward migration. In China and North America the mountain ranges run north—south, and forest species, retreating before the glacial advance, were able to reach warmer latitudes.

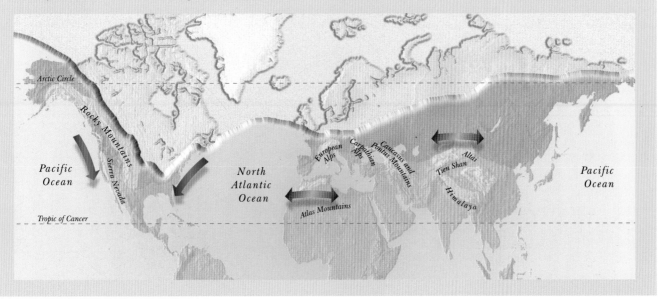

life on earth. Magnolias are among the most primitive of flowering plants; fossils have been found in rocks from the Cretaceous period (144-65 million years ago). Their simple shape, thick petals, and rigid anthers, together with a powerful scent are adapted to pollination by beetles, whose blundering flight may be similar to that of the earliest insects.

Southern belle

Magnolia grandiflora, from the river valleys of southeastern America, grows in warm climates as a freestanding tree, and is a familiar street tree in California. It soon became popular in northern Europe following its introduction some time before 1730, but welcomed the protection of a warm wall. The seedlings, as with many magnolias, were slow to flower, and it was not until August 1737 that the first plant actually bloomed in Europe at the home of Admiral Sir Charles Wager (1666-1743), First Lord of the Admiralty, in London.

During the summer, large creamy-white flowers nestle among evergreen, russet-backed foliage, like "great white pigeons settling among dark leaves" as the English writer and gardener Vita Sackville-West described the magnolias in her Sissinghurst garden. The fragrance they release is strong and lemony; the native Americans avoided sleeping beneath a flowering tree, believing that the scent would overpower them. A variety of *M. grandiflora* is 'Little Gem', which is a slow growing form that flowers when it is only 3 feet (1m) tall.

Oriental hardies

A long history of cultural and horticultural exchange between China and Japan often blurred the wild origins of the plants which Europeans were acquiring at the treaty ports. Many Japanese species first reached Europe from China, and others, like *Magnolia denudata*, wild in central China, traveled via Japan. The fragrant flowers of *M. denudata*, which, as the name implies, appear on bare branches in early spring, were regarded by the Buddhists of the Tang dynasty (A.D. 618-906) as symbols of purity, and the shrub was widely planted in temple gardens. The flowers were also often depicted on porcelain and in paintings, and their petals, dipped in flour and fried, are still regarded as a delicacy. Buddhism, and the *Yulan* (lily tree) as the Chinese call *M. denudata*, reached Japan during the Tang period, and from there the British botanist Sir Joseph Banks

. . . .

FRENCH INNOVATION
M. x soulangiana (right)
is the most widely grown of
all magnolias. In favorable
seasons the flowers may be
followed by strange,
knobbly fruits containing
orange and black seeds.

Master of botanical art

The German botanical artist Georg Ehret (1708-1770) recorded the first flowering of *Magnolia grandiflora* in the summer of 1737, walking 3 miles (5km) each way to the gardens of Admiral Sir Charles Wager on the outskirts of London. Ehret was born in humble circumstances in Heidelberg, Germany, and was taught to draw by his father. He continued to draw when he later became an apprentice gardener. In 1733, while in Holland, he met Carolus Linnaeus (1707-1778), father of modern botany, from whom he learned of the new artificial classification of plants. This training stood him in good stead in later years, when he was commissioned to illustrate several important scientific works, including Linnaeus' *Hortus Cliffortianus* of 1737, a record of the plants grown in a Dutch garden. Ehret moved to London in 1736, at a time when flower painting was fashionable, and found himself in demand as a teacher and illustrator. The 20th-century historian of botanical art, Wilfrid Blunt, described Ehret's work as a "fine compromise" between art and science, capturing both the spirit and the detail of the plants he painted.

introduced the first specimen to Britain in 1789. Although the tree is extremely resistant to cold, the early flowers of *M. denudata* are easily damaged by frost, a problem avoided by the later flowering *M. liliiflora*, another Chinese species first introduced to Europe from Japan in 1790 by the Duke of Portland. The flowers of the *Mulan* (woody orchid) are flushed purple on the outer surface, and white within, but can be rather untidy in appearance.

French experiments

Chevalier Etienne Soulange-Bodin (1774-1846) was sickened by his experiences in the Napoleonic Wars, so he took up gardening. He wrote in 1819: "The rising taste for gardening becomes one of the most agreeable guarantees of the repose of the world."

Soulange-Bodin became the founding director of the Royal Institute of Horticulture based at Fromont, near Paris. In what must have been a very early experiment at the Institute, *M. denudata* was crossed with *M. liliiflora*, for by 1827 the new hybrid was being praised for its beauty. The flowers coupled the refined shape of *M. denudata* with the color and later flowering of *M. liliiflora*, although like *M. denudata*, they are borne on leafless branches. The hybrid was named *Magnolia x soulangiana*, in honor of the Chevalier and has since become the most widely planted and popular magnolia throughout Europe and North America. Over a hundred distinct cultivars have since been created. Most familiar are those with the flowers stained pinkish-purple on the outside and white internally, but there are pure white and dark pink clones as well. *M. x soulangiana* is considered hardy to -20°F (-29°C), and flowers in early spring, often covering the otherwise naked tree with innumerable flowers.

Magnolia grandiflora by Ehret

"WOODY ORCHID"
Magnolia liliiflora (left),
known in Japan as Mulan,
meaning "woody orchid",
is a native to China, but
reached Europe via Japan,
where it has been
cultivated for centuries.

$\mathscr{S}$OME OF THE OLDEST LANDSCAPED GARDENS IN NORTH AMERICA ARE IN THE STATE OF SOUTH CAROLINA, WHERE WEALTHY SETTLERS BUILT PROSPEROUS PLANTATIONS AND ELEGANT GARDENS TO GO WITH THEM.

$\mathscr{G}$ARDENS OF THE $\mathscr{S}$OUTHERN $\mathscr{P}$LANTATIONS

In the 1700s, wealthy settlers from Europe established themselves in the southern parts of America. They established plantations of cotton, indigo, tobacco, and rice, and created ornamental gardens which were a true blend of Old and New World styles. The grand Classical formality of late 17th- and 18th-century Europe was softened by a naturalistic approach favored by the American settlers, which adapted to the local environment and could be managed alongside the plantation crops.

The mild winters, hot summers, fertile soil, and moderate rainfall of South Carolina supports a rich variety of native plants, including the stately live oaks (*Quercus virginiana*), often draped with festoons of grey Spanish moss, and swamp cypresses, which provide dramatic framework and focal points.

Middleton Place, near Charleston, South Carolina, dates from 1741—although it took 100 slaves nearly a decade to complete the landscaping—and is one of the oldest landscaped gardens in North America. Inspiration came from France, the garden's formal symmetry and precise pattern are reminiscent of the architectural gardenscapes of Andre le Notre, who masterminded the French King Louis XIV's gardens at the Palace of Versailles. Water and woodland were key features of Le Notre style, and ideally suited to Middleton, located as it is, on a bluff above the river and the marshy Carolina Low Country. But over the basic formality was superimposed a naturalistic style that was characteristically American Colonial. It was a direction later emulated by the presidents George Washington, at Mount Vernon, and Thomas Jefferson, at his hilltop Palladian villa at Monticello. The Middleton gardener was English, and the plants were a mixture of indigenous and exotic. The second Henry Middleton was a friend of the great French botanist André Michaux, who brought a host of newly discovered plants from Europe and the Far East to South Carolina. Among them were, in 1786, the first *Camellia japonica* to be planted in an American garden. One of the original camellias survives still at a corner of the parterre, but many other varieties have since been introduced and line the allées of the formal garden. Henry's son, William, introduced *Azalea indica* (syn. *Rhododendron indica*), which were supplemented in the 20th century by the brilliantly colored Kurume azaleas introduced to the West from China by Ernest H. Wilson. The blend of formal and the natural is echoed in many other plantation layouts and in the smaller walled gardens of South Carolina.

1. *Reflection Pool*
2. *Ruins of main house*
3. *Parterre, terraces, Butterfly Lake*
4. *Octagonal Garden*
5. *Camellia allées*
6. *Sundial Garden*
7. *Azalea pool*

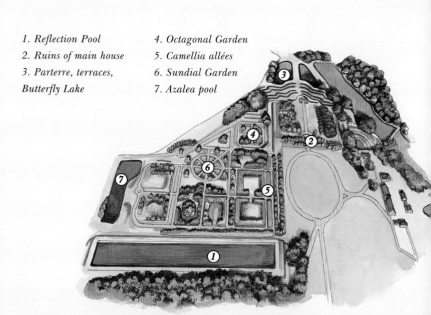

MIDDLETON PLACE
GARDEN (*above and left*) *is
planned, in typical
Classical 18th-century
style, around a central,
uninterrupted vista from
the front of the house. A
series of curved terraces
once doubled as paddy
fields which were
alternately flooded and
drained. A pathway
between ornamental lakes,
leads to the river and*
*marshland beyond. The
formal garden northwest of
the house was planned
with great precision. Near
to the house is a series of
small gardens, including
the Octagonal Garden
where the gentlemen used
to play bowls while the
ladies watched from the
terraces, and the Sundial
Garden which was planted
with roses in the 18th and
19th centuries.*

FORMAL "ALLEES"
*(right), typical of formal
French style, lead to secret
arbors and surprise
viewpoints. This one is
carpeted with* <u>Camellia
japonica</u> *petals. The first
camellias were introduced
to Middleton in the late
1700s.*

"$\mathcal{L}$AY HER I' THE EARTH; AND
FROM HER FAIR AND UNPOLLUTED
FLESH LET VIOLETS SPRING..." FLOWERS WITH ADDED VALUE, WHOSE
NATURAL MODESTY IN THE WILD HAS LED TO ASSOCIATIONS WITH
THOUGHTFULNESS, SHYNESS, REMEMBRANCE, AND HEALING,

Viola 'Belmont Blue'

$\mathcal{P}$ANSIES AND $\mathcal{F}$ORGET-ME-NOTS
(VIOLACEAE & BORAGINACEAE)

The large-flowered cultivated pansy of today is derived from a simple weed common in European cornfields, *Viola tricolor*. The flowers of *V. tricolor*, or "heart's-ease," as it is known, are no more than 1 inch (2.5cm) long, with a yellow lip, a pair of whitish "wings" and purple upper petals, which usually have fine dark lines converging at the center. Many variants occur in lighter or darker shades; the darkest being named as 'Bowles' Black', after the great gardener who first distributed it in Britain before the First World War. It is a velvety

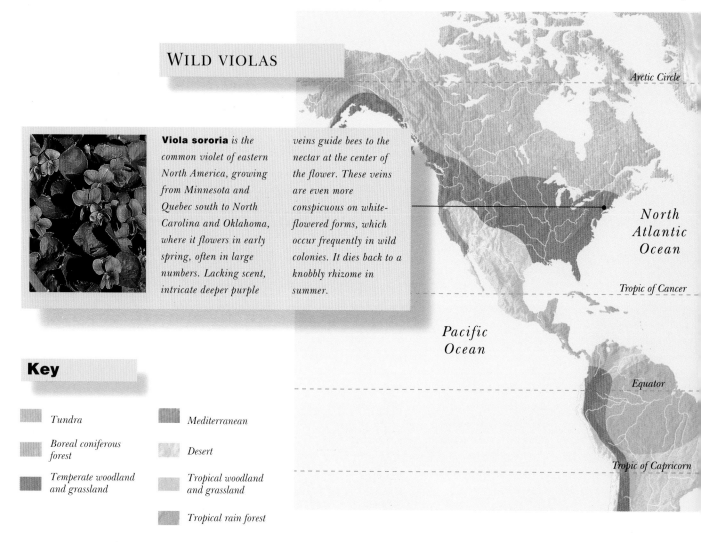

WILD VIOLAS

Viola sororia *is the common violet of eastern North America, growing from Minnesota and Quebec south to North Carolina and Oklahoma, where it flowers in early spring, often in large numbers. Lacking scent, intricate deeper purple veins guide bees to the nectar at the center of the flower. These veins are even more conspicuous on white-flowered forms, which occur frequently in wild colonies. It dies back to a knobbly rhizome in summer.*

Arctic Circle

North Atlantic Ocean

Tropic of Cancer

Pacific Ocean

Equator

Tropic of Capricorn

Key

Tundra	Mediterranean
Boreal coniferous forest	Desert
Temperate woodland and grassland	Tropical woodland and grassland
	Tropical rain forest

purple-black, with a yellow eye, and like all forms of the annual *V. tricolor* it self-sows prolifically. The transition from heart's-ease to pansy only began in the early 1800s.

Although several British gardeners started experimenting at about the same time, it is generally agreed that the father of the pansy is William Thompson, gardener to Lord Gambier

at Iver in Buckinghamshire. Lord Gambier (1756-1833), known to his contemporaries as "Dismal Jimmy," was an admiral in Lord Nelson's navy, but he retired in 1811 following a court-martial. Perhaps to occupy his mind, he seems to have taken an interest in gardening.

In 1813 he brought Thompson some plants of *V. tricolor* which he had found on the estate, and suggested that he try to "improve" them. At first Thompson bred for size and colors by simple selection, but he later began to hybridize *V. tricolor* with other species, including the yellow-flowered European mountain plant, *V. lutea* and the purple Siberian *V. altaica*, which are both perennials. These species contributed different colors and a more compact habit to the pansy, while Thompson and other breeders aimed to produce rounder flowers.

The characteristic dark blotches that make up the "face" in the center of a modern pansy appeared by chance, in about 1830, on a self-sown seedling in a neglected corner of Thompson's domain. Its value was immediately

Viola tricolor *was the pansy known to the poet William Shakespeare as a symbol of thoughtfulness, the name being derived from the French* pensée

("thought") Found wild in arable fields in northern Europe, its pretty "face" and soft colors have endeared it to gardeners for centuries, even though it produces abundant seedlings which may have to be weeded out. Many color forms have been selected, ranging from creamy-yellow to purple-black.

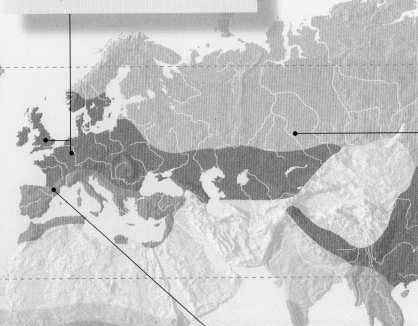

Viola altaica *is a small pansy from the mountains of Siberia, including the Altai Mountains, where it flowers in summer. Introduced to Europe as early as 1805, it was a*

parent of the familiar garden pansies, contributing its stocky habit and purple color to the lanky, paler pansies of the early 1800s. Now very rare in cultivation, it deserves to be more widely grown, although it probably needs careful attention to flourish.

Viola cornuta *is an alpine pansy from the Pyrenees of Spain and France. Its name, which means "horned," is derived from the long, nectar-containing spur behind the flower. Forming low mats, it is a good plant for the rock* *garden or front of border, and was a parent of the perennial "tufted violas", which can be planted in similar situations. The viola is easily raised from seed to flower in the first year.*

COLOR CODE
The three-colored flowers of the wild pansy, or heart's-ease (left), gave it the scientific name Viola tricolor.

appreciated, and from that chance discovery all blotched pansies are descended.

Waxing and waning popularity

In Britain the early pansy became a florist's flower (*see p.51*), with breeders striving for rounded, show-quality blooms and regular markings. However, continental European growers favored a much less formal look, and it was not long before their richly colored flowers became popular throughout the world, superseding the show pansies. The pansy color palette has widened over the last century to include pure orange and pale pink, often lacking the black blotch. Many strains will now flower throughout the winter, making the plants ideal for winter bedding

While pansies (*Viola* x *wittrockiana*) were being improved during the late 1800s, a parallel development, using the Pyrenean *Viola cornuta*, led to the creation of the violas or tufted pansies. These are similar to pansies, but have smaller, often unmarked flowers, and are true perennials that can be easily propagated from cuttings. Although most are not widely grown today, a few, such as the greenish brown 'Irish Molly' and the red and yellow 'Jackanapes', are popular as edge-of-border plants.

Scented violets

The 500 species in the genus *Viola* divide into those that can be called pansies, usually with flattened flowers in shades of blue and yellow, and those known as violets, with more concave flowers in shades of purple. The garden violet is

TIDY HABITS
The garden pansy (above) is rounder faced and more compact in habit than its wild relations.

derived primarily from the European sweet violet, *V. odorata*, a plant of woodland edges that flowers in early spring before the trees come into leaf. Typically dark violet, its color varies considerably, and white forms are not uncommon. In the 18th and 19th centuries *V. odorata* was hybridized with *V. alba* and *V. suavis* from eastern Europe and Russia. This created scented violets in pale blue and even pink and crimson. Hybridizing with the American *V. sororia*, with its white or mauve flowers veined blue-black, produced large flowers on vigorous plants, but lost the scent.

William Shakespeare, whose observation of natural history was acute, described violets as "Stealing and giving odor." And indeed, if you smell a violet, you will find that, after a short time, the scent seems to have vanished; sniffing it again in a minute or so, you will find that it has "returned." The explanation is simple; the chemical constituent of the perfume that gives the characteristic scent, ionine, clogs the scent receptors in the nose, preventing the detection of the scent. If you turn away from the flower, the receptors unblock, and the scent is once again detectable.

Violets were the favorite flower of the French Emperor Napoleon Bonaparte (1769-1821). As a token of his love for his wife Joséphine (1763-1814), he asked the gardener of her estate, La Malmaison, to send her a bunch every year on their wedding anniversary. On being exiled to Elba in 1814, he told his supporters that he would, "return with the violets in the spring." He kept this promise by escaping from Elba in March 1815—the start of the "Hundred Days" that ended at the Battle of Waterloo. However, Josephine had died during his exile. Bonaparte picked and pressed some violets from her grave and wore them in a locket until his own death in exile on the island of St. Helena.

MYOSOTIS: FORGET-ME-NOT
The common name of "forget-me-not" originally described *Myosotis scorpioides*, the water forget-me-not of stream banks and marshes, but it later referred to all members of the genus. The name was used in medieval France (*ne m'oubliez mye*) and Germany (*vergiss-mein-nicht*), but its origins are uncertain, long predating the German legend that tells of a knight who was swept away by a stream when picking the flowers for his lady. As he disappeared he cried, "forget me not" and flung

the flowers to her. Oddly, in Britain, the name seems to have been used until the early 1800s for the blue-flowered speedwell, *Veronica chamaedrys*; *Myosotis* was known as scorpion grass, because of its coiled flower stems, resembling a scorpion's tail, which straighten as the flowers open.

The sky-blue, five-lobed flowers of forget-me-nots have a contrasting white or yellow "eye" and often open from pink buds. The color contrast caught the attention of the German botanist C.P.J. Sprengel (1766-1833), who reasoned that it must serve some purpose and

deduced that contrasting colors in flowers were signals to attract pollinating insects. In some cultivars of the commonly grown bedding forget-me-not, *M. sylvatica*, the flowers remain pink, while others are pure white. *M. sylvatica* is a European plant, but other species are found throughout Asia and North America, and several occur in the mountains of New Zealand.

New Zealand giant

Chatham Island is a speck of land lying several hundred miles east of New Zealand's South Island. Most people have never heard of it, but

RARE FRAGRANCE
Viola odorata (left) is the deliciously scented sweet violet, a wild plant of hedgerows and woodlands across Europe, and one of very few fragrant species.

PLANTHUNTING IN
NORTH AMERICA
*By the 1750s, North
America was established as
a significant source of new
plants ripe for cultivation.
The map (right) highlights
some key expeditions.*

SHOW STOPPERS
*Gigantic echiums (left)
from the Canary Islands
make a spectacular show
in frost-free gardens,
where they bloom in late
spring.*

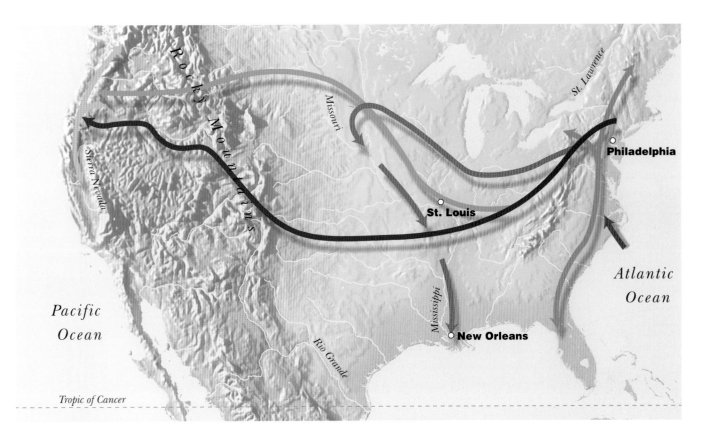

to gardeners it is famous as the home of the giant forget-me-not, *Myosotidium hortensia*. Although a member of the Boraginaceae family, *Myosotidium* is not closely related to *Myosotis*. However, its rounded blue flowers are similar enough to justify its common name. Borne in heads above large, glossy green leaves, the flowers are actually two-toned, being darker at their center than at the edges. When it was introduced to Europe in 1846, gardeners were convinced that it would only thrive with a mulch of seaweed and rotting fish. Fortunately, however, a fertile, moisture-retentive compost is quite acceptable. Best grown in mild, humid locations, *Myosotidium* needs protection from prolonged frost. Sadly, the numerous sheep on its native island find the giant forget-me-not very palatable, and it can now only be found on cliffs out of their reach.

Atlantic isolation

Most members of the borage family are low-growing herbs and annuals, often with rather harshly hairy leaves, and flowers that turn blue from pink, although the family does contain trees and shrubs. A typical genus is *Echium*, which in continental Europe is represented by several short-lived herbs, including *E. vulgare*, "vipers' bugloss." This forms prostrate rosettes of leaves from a spire of blue flowers, up to 3 feet

(92cm) high, which emerges in its second year. Transported to America in its European role as an ornamental, or bee-attracting plant, it has become a serious weed.

Much more exciting are its relations from the Canary Islands and Madeira. Isolated in the Atlantic Ocean, *Echium* species have evolved in quite a different direction, and become shrubby or gigantic. The largest is probably *E. pininana*, which is only found on La Palma in the Canary Isles. Its inflorescences often reach 12 feet (3.6m) in height and are composed of thousands of blue flowers and pink buds—an amazing spectacle when in full flower. *E. pininana* and the related *E. wildprettii* take several years to flower, building up a large rosette before the inflorescence develops. Once the seed is ripe the plants die. Unlike *E. pininana*, which has a rosette borne on a stem, the silvery rosette of *E. wildprettii* hugs the ground. The inflorescence reaches 6 feet (1.8m) or more, with coral-pink flowers among long gray bracts. It is a native of the volcanic slopes of Tenerife, and said to be threatened in the wild. Fortunately, it is easily cultivated and, being a much smaller plant than *E. pininana*, can be conveniently pot grown where winters are frosty. These spectacular plants, as well as the shrubby *E. candicans*, the "Pride of Madeira," are at their best in a Mediterranean or maritime climate.

*T*HE WORLD'S CLIMATES PROVIDE
A WEALTH OF PLANTS STRONG ON
FORM AND LEAF STYLE, FROM DELICATE FERNS TO SCULPTURAL SHRUBS
AND TREES, GROWN FOR ARCHITECTURAL AND BACKGROUND EFFECTS.

Gunnera manicata

*F*OLIAGE *P*LANTS

*P*lants grown for their leaves, texture, structural form, and overall "architecture" have been popular for centuries in Oriental gardens. They come from a huge range of families, from simple and primitive mosses and ferns, to highly evolved flowering plants such as yuccas and agaves. In recent times, they have contributed increasingly to Western planting schemes. Plants like bamboos and reeds, grasses and sedges are no longer the "Cinderellas" of

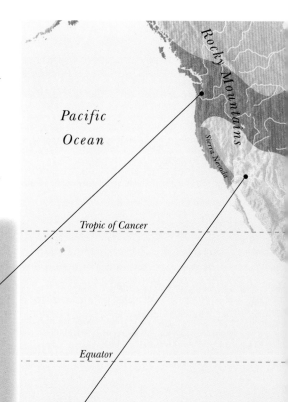

*Pacific
Ocean*

Rocky Mountains

Sierra Nevada

Tropic of Cancer

Equator

Polystichum munitum, *the "western shield fern", grows in the damp forests of western North America from Mexico to Alaska. It is one of many species of shield fern from cool places around the world, most of which are handsome foliage plants* *for the woodland garden. Like most ferns,* Polystichum *species require high humidity for successful cultivation.*

Yucca gloriosa, *from the southwestern U.S., is a robust member of a drought-tolerant genus. Although principally grown for its rosettes of sharp-tipped leaves, which have given it the English namd "Spanish Dagger," it does often bear large inflorescences* *of scented white flowers. The fleshy, wax-covered leaves are a water-saving adaptation to the dry conditions of their natural habitats.*

FOLIAGE FROM ALL CLIMATES
Examples from different vegetation zones in the Amercas show how adaptations to climate affect leaf shape and plant architecture.

cultivation but artfully used to form a strong framework, an architectural feature, complementary backdrop, or an attractive and useful groundcover.

GRASSES AND SEDGES

In the West and temperate regions, the traditional use of grass is for lawns. But many members of the grass family, Gramineae, are effective feature plants in their own right. One of the most widely cultivated is *Milium effusum* 'Aureum', derived from a European woodland species, with soft golden-yellow leaves suited to the light shade beneath trees. It was first spotted early this century in the Birmingham Botanic Garden, by the plantsman E. A. Bowles (1865-

1954), hence one of its common names, Bowles' golden grass. Bowles grew an enormous diversity of species in his garden at Myddelton House, near London and championed the cultivation of grasses at a time when they were not popular as ornamentals. He is often regarded as an expert on bulbous plants, but his "eye" for any curiosity is recalled by the many plants named after him, and he also wrote about his work in a charming and deeply informative manner.

Many grasses and sedges (*Carex* spp.), with their open inflorescences and loose foliage are used to soften hard landscaping or tone down bright border plants. *Miscanthus sinensis*, a clump-forming grass from eastern Asia, can reach 10 feet (3m) in height, although many of its cultivars grow lower and are more suited to small gardens. Several cultivars are variegated, including 'Zebrinus', with white-banded leaves, and the yellow-banded 'Strictus'. The striped leaves bristling from the clump has led to the nickname of "porcupine grass." The plume-like inflorescences of *Miscanthus* appear in late

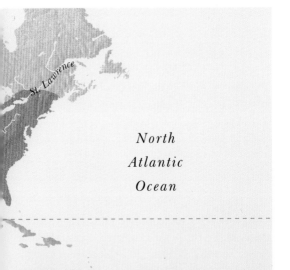

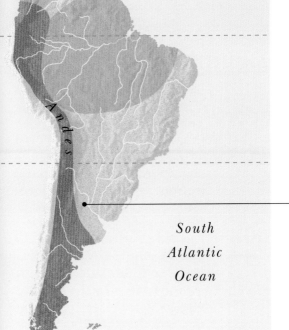

Monstera deliciosa, *a member of the Arum family (Araceae) from the forests of Central America, has large* leaves adapted to capture light in the gloom of the forest understory. Their elongated apex allows rain to run off easily. The function of the large holes, which give the plant the common name "Swiss cheese plant," is unknown. Creamy-white leaves are followed by edible fruit.

Key

	Tundra
	Boreal coniferous forest
	Temperate woodland and grassland
	Mediterranean
	Desert
	Tropical woodland and grassland
	Tropical rain forest

Cortaderia selloana *is a large, clump-forming grass that covers large areas of the plains of South America—the Pampas—which gives it the name "Pampas grass." Its long leaves have a razor-sharp margin which deter razing animals. In* cultivation, pampa grass is used as a feature plant being particularly impressive when it produces its plume-like inflorescences.

North Atlantic Ocean

St. Lawrence

Andes

South Atlantic Ocean

GOLDEN WONDER
The bright yellow growth of _Milium effusum_ 'Aureum' (left) is particularly effective when planted near dark green foliage.

Streaks and speckles

Variegated plants are usually natural variants, and not diseased, as many people think. The effect is due to patches of microscopic cells which do not produce the normal green pigment, chlorophyll. The pattern of pale areas depends on the locations of these cells in the developing bud. It may take the form of longitudinal streaks, central blotches or irregular speckles. Variegated specimens can make a fine contrast planted among non-variegated plants, but take care with siting them, since many are less robust than their fully-colored relations.

summer, and if not zealously tidied in fall, dry brown stems remain as a winter feature.

Several New Zealand sedges, such as *Carex comans* and *C. buchananii*, have naturally brown leaves and their tousled clumps provide year-round color. Another colorful grass, *Phalaris arundinacea* 'Picta', is also known "gardener's garters." It appeared in England during the 1500s and has bright green and white foliage. It needs to be sited with great care, since it can be extremely invasive.

Largescale groundcover
The grassy plains, or *Pampas*, of southern South America share their name with the dominant plant that grows there, *Cortaderia selloana*, which covers large areas of Argentina and southern Brazil. Its large clumps of sword-shaped evergreen leaves sprout white plumes in late summer. The leaves are edged with a razor-sharp margin of silica (the mineral constituent of sand), and Spanish settlers dubbed it *cortadera*, meaning "the cutter". This grass requires good soil, full sun, and winters that are not too harsh. With such conditions it can then withstand exposed places, even salt spray.

The giant reed, *Arundo donax*, grows profusely in damp places around the eastern Mediterranean. It is one of the largest herbaceous grasses, reaching 20 feet (6m), and its stems arch gracefully under the weight of glaucous leaves, which sway

RAPID GROWER
The giant reed _Arundo donax_ (below) achieves heights of up to 20 feet (over 6m) within one season, flourishing especially in warm, wet conditions.

and rustle in the slightest breeze. It can be grown as a greenhouse plant in northern countries and, like bamboo, provides rigid canes, reeds for woodwind instruments, and hollow stems for making into Pan Pipes.

PHORMIUM
The New Zealand flaxes—*Phormium tenax*, *P cookianum*, and their many cultivars with their large, spiky tufts—are often planted where a strong contrast of form and color are required. Their common name refers to the strong fibers which can be extracted from the sword-shaped leaves. The native Maori people not only used the fibers to make clothing, nets, and rope, but also appreciated the plant's decorative qualities, and had selected several variegated clones which they grew around their homesteads long before the first Europeans arrived.

Variegated and red-flushed forms occur occasionally in the wild, but most of the brightly colored cultivars available today are the products of hybridization and selection in Europe and North America. An original Maori cultivar is *P. tenax* 'Radiance', which was first recorded in 1870. It has bright green leaves striped with yellow, as well as a reddish margin. All phormiums are striking plants, and as different cultivars range in size from dwarfs less than 1 foot (30cm) in height to giants of 10 feet (3m) or

more, there is a cultivar for all sizes of plot. They are very tolerant of strong wind and salt spray, and suitable for seaside planting, but most are not hardy below 13°F (-10°F).

BAMBOOS

The 1,000-plus species of bamboo, grasses with woody stems (culms) and stalked leaves, are widely distributed in tropical and warm parts of the world, most abundantly in eastern Asia. They are immensely practical plants, and used for a multitude of purposes—construction, decoration, food. They have been graceful elements in the gardens of China and Japan for more than 2,000 years.

Hardy bamboos were introduced to Europe in 1823 and rapidly became popular as specimen plants, in rows as screens, or in clumps to give a jungle-like effect. The *Bambouserai de Prafrance*, in southern France, where many species are grown in a natural setting, houses one of the finest collections in the world.

Fargesia murieliae, which forms elegant clumps of medium height, with arching culms

NON-INVASIVE

Bamboo (above) can contribute Oriental elegance especially if it contrasts with bright colors and stiffer shapes nearby. Choose clump-forming species such as Fargesia murieliae which will not overrun the garden.

bearing small leaves, is suited to small plots in temperate regions. It was introduced from western China by Ernest Wilson in 1913, and named after his daughter, Muriel. With a similar species, *F. nitida*, it forms almost impenetrable thickets in its natural mountain habitat, and forms part of the diet of that very rare mammal, the giant panda.

One of the remarkable aspects of bamboo biology is the simultaneous mass flowering of plants over a wide area, followed by the death of the woody culms. The plants regenerate naturally, from either rhizomes or seeds, but during the interval, the pandas lack food and are forced to move on. In the past, this caused few problems. But the population of China has expanded so greatly that suitable bamboo is rare, and when it does flower and die, the pandas cannot easily move elsewhere. The cause of the mass flowering is still not fully understood. It appears to occur when the plants achieve a certain physiological state which triggers flowering, and is perhaps related to age and climate.

EUPHORBIA

Euphorbia robbiae is a widely grown groundcover plant, whose natural habitat of dry eastern Mediterranean forests makes it suitable for "difficult" dry corners. Its whorls of dark green evergreen foliage are enlivened by yellow-green inflorescences in spring. The species was introduced to Europe in 1891 from Turkey, by Mrs. Mary Anne Robb of Hampshire, England. She had attended a wedding in the region, and as a keen plant collector, she also seized the opportunity to acquire a new plant. She commanded her guide to dig up the plant and then, without a specialist container, she kept it in her hat box. What happened to the imposing hat, which was intended to impress Turkish officials by its magnificence, is unrecorded. But the plant has been known as "Mrs. Robb's Bonnet" ever since.

Other euphorbias, *E. characias* and its subspecies *wulfenii*, form shrubby growths with dense whorls of glaucous leaves and large, terminal, green and yellow inflorescences. They thrive in well-drained conditions in full sun, imitating their natural habitat of Mediterranean hillsides. The familiar Christmas poinsettia is *E. pulcherrima*. At first sight, this native of western Mexico does not resemble the herbaceous euphorbias of north temperate regions. But it shares the same curious flower structure and the

BROAD LEAVES
The broad, fleshy leaves of hostas provide low-level sculptural effects, especially when different cultivars in a range of green, blue-green, gold and variegated forms are combined.

white latex found in all members of the genus—which is named after Euphorbius, a Greek physician of the first century A.D. He was medical advisor to King Juba II of Mauritania, and apparently used the sap for medicinal purposes. This is not advised today, for it is considered poisonous, and causes a nasty rash on sensitive skin.

The poinsettia is both artificially dwarfed and forced into flowering for the goodwill season. Most specimens then promptly die, but in a frost-free situation, they can grow as magnificent shrubs covered in scarlet bracts.

HOSTAS: PLANTAIN LILIES

Hostas or plantain lilies, like day lilies, are indigenous to Japan and China. They arrived in Europe in the 1800s and are now valued for their bold effect in gardens, provided by their foliage rather than the flowers. Many cultivars of *Hosta sieboldiana* have large leaves with a strong bluish sheen, although others are golden or variegated. A smaller species, *H. undulata*, has twisted leaves, which are also variegated in clones such as 'Univittata'.

All hostas enjoy light shade in fertile or moist soil. They are useful summer groundcover plants, becoming dormant in winter. Most produce attractive spikes of flowers, usually pale purple or white, and in some species, such as *H. plantaginea* the blooms are fragrant. However, slugs and snails love to eat hosta foliage, and they can quickly turn it into unsightly tatters.

YUCCAS AND AGAVES

Several species of shrubby, lance-leaved *Yucca* have variegated forms, which make them effective foliage plants for hot, dry

FLUORESCENCE
The dazzling yellow-green flowerheads and bluish-gray foliage make Euphorbia characias *(left) a bright addition to a prominent, sunny site which reflects its Mediterranean origins.*

A craze for ferns

In the mid and late 1800s, ferns became exceedingly popular house plants in Europe, especially Britain, and in the U.S. Most tolerate low light levels and a humid atmosphere. Such was the demand that some localities lost almost of all their natural ferns, and some rarer species never recovered from these depredations.

Most prized were those with finely divided fronds, dissections or crests at the tips of the pinnae (individual front "blades"). Some of these mutations were extremely beautiful and are still widely grown. Among the most remarkable is *Athyrium filix-femina* 'Victoriae' named after Queen Victoria. Its pinnae form a criss-cross pattern, each bearing a small crest at the end, while the main frond-tip itself bears a larger tassel. It was discovered in Scotland in 1861 by James Cosh, a student. He jumped over a wall into a field, narrowly missing the fern—which he then noticed and dug up.

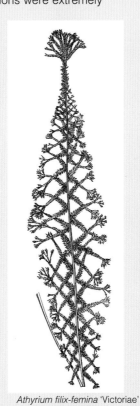

Athyrium filix-femina 'Victoriae'

Another lucky find was *Polystichum setiferum* 'Pulcherrimum Bevis', which was discovered in 1876 in a ditch in southern England, by a laborer, John Bevis. Similar mutants are also found in tropical ferns. Examples include the plumose cultivars of *Nephrolepis exaltata,* suited to warm gardens or greenhouse cultivation.

conditions. This is an entirely American genus, first reaching Europe in the late 1500s, when the early specimens were hugely prized. In 1629, John Parkinson described how he came by a plant descended from one owned by John Gerard, who had named the genus—but in error. Gerard had heard of a South American plant called *yuca*, which is the Spanish name for cassava or manioc, *Manihot esculenta*, also known as the "tapioca plant". Its starchy, fleshy tuber is an important food crop throughout the tropics. Gerard's mistake was rapidly realized, but the name *Yucca* stuck.

Gerard had sent a plant to his friend Jean Robin (1550-1629), a nurseryman in Paris and the French royal botanist, caring for the gardens of the Louvre in Paris. His son Vespasian Robin (1579-1662) sent a piece of the yucca to Sir John de Franqueville in England, who passed it on to Parkinson. Yucca leaves are typically in rosettes and each leaf is slim, stiff and sword-shaped, the point tipped with a very sharp spine: careful siting is desirable to avoid pricked skin and poked-out eyes. The rosette grows for several years before flowering, which may be irregular, but which then produces dramatic inflorescences of scented, waxy-white bells. *Y. gloriosa* 'Variegata,' a native of southeastern North America, is hardy enough for cultivation in temperate regions.

The agaves or "century plants" of the genus *Agave*, resemble the yuccas with their rosettes of long, slim, sharp-tipped leaves. And they also grow for many years—although not usually a century—before producing the massive inflorescence which in some species may be 30 feet (10m) long. The bloom develops rapidly using the stored starches or carbohydrates in the leaf rosette. In Mexico, the developing flowers' sugary sap is tapped, fermented, and distilled to produce the spirit tequila.

Most agaves are natives of dry places in the Americas and Caribbean. One of the most popular ornamental species is *A. americana* 'Variegata', with broad creamy-yellow stripes along the edges of the grey leaves.

The largest leaves

The largest leaves of any plant that can be grown outdoors in temperate zones belong to *Gunnera manicata* (Gunneraceae). The name commemorates Ernst Gunnerus, a Norwegian bishop in the 1700s. Its huge, prickly, umbrella-like leaves may be 10 feet (3m) across and 5 feet (1.5m) high. This species comes from the damp,

mild forests of Chile and needs ample moisture during the summer growing season. So it is usually grown by water, and in a sheltered location, since the plant is winter hardy, but the leaves are frost sensitive.

CONIFERS

Conifers are gymnosperms, literally, "naked-seeds," more ancient than the flowering plants, or angiosperms. They produce inconspicuous flowers—although their fruits, the cones—are often large and handsome.

One of this century's most widely planted trees is the columnar Leyland cypress, x *Cupressocyparis leylandii*. It is a hybrid between species from two distinct genera: *Chamaecyparis nootkatensis*, the Nootka Cypress from the northwestern coast of North America; and *Cupressus macrocarpa*, the Monterey cypress from California. The Leyland combines features of both parents into a hardy, fast-growing tree, suitable as a large specimen or for hedging.

The cross leading to x *Cupressocyparis leylandii* first arose as chance seedlings in an English arboretum in 1888. It has been repeated on many occasions, producing numerous cultivars with different characteristics of habit and color. 'Castlewellan' foliage is gold in summer, bronze in winter, and tolerates atmospheric pollution and salt spray.

Dwarf versions

Many dwarf conifers grown for their low maintenance and all-year foliage have been selected from wild populations with certain features. For example, the Irish juniper, *Juniperus communis* 'Hibernica', was derived from a wild colony with a rigidly erect habit, rather than the typical juniper's sprawling bush shape, and can contribute a useful vertical element to a planting scheme.

Similarly dense and compact is the spruce *Picea glauca* var. *albertiana* 'Conica', which forms a thick, conical bush. It was spotted originally in the wild in 1904, by J. G. Jack of the Arnold Arboretum, Boston, Massachusetts. While crossing the Rocky Mountains by railroad in Alberta, Canada, he noticed the plant through the window, pulled the communication cord, and nipped out to collected propagation material while the train was stopped. His act

LOCAL SHADE *The huge leaves of Gunnera manicata (left) unfurl like umbrellas in late spring; the bushy inflorescences remain beneath them.*

Tradescant tradition

The American genus *Tradescantia* (Commelinaceae) is named for the Tradescants, father and son, gardeners to King Charles I of England. Both were greatly interested in America, and they became the pivot of a network of gardening friends spanning both Europe and North America. The father John (?-1638) made at least seven overseas trips, to Russia, Spain, France and the Low Countries, but never actually crossed the Atlantic. John the Younger (1608-1662), visited Virginia at least three times, and is credited with the introduction to Europe of the North American species *Aquilegia canadensis*, *Aster tradescantii*, and *Rudbeckia laciniata,* as well as the spiderwort, and many trees and shrubs.

Tradescantia virginiana, by Redouté, 1805.

CHILE DISCOVERY
The monkey puzzle tree,
<u>Araucaria</u> <u>araucana</u> *(left)*
was discovered by
Archibald Menzies (1754-
1842), surgeon and
botanist on Captain
George Vancouver's
exploration of the western
coast of the Americas
1792-95.

may have inconvenienced his fellow passengers, but it greatly benefitted gardeners!

Large specimens

Several species of conifer are grown as majestic specimen trees in very large gardens and parks. They include the cedar of Lebanon, *Cedrus libani*, with its umbrella-like form of spreading branches and thick trunk; the related Atlas mountains cedar, *C. atlantica*, the blue-leaved 'Glauca' being very popular for its coloration and less widely spread branches; the Himalayan cedar, *C. deodara*; and the Chilean pine or monkey puzzle tree, *Araucaria araucana*, a native of the dry volcanic slopes of the Andes in Chile and Argentina. The common English name was inspired by a visitor to the Cornish garden where one of the first saplings was growing, around 1850, and who stated: "It would puzzle a monkey to climb that tree."

ANCIENT MAJESTY *The stately Cedar of Lebanon,* <u>Cedrus</u> <u>libani</u> *(above), from the mountains of southern Turkey and Lebanon, has been admired since Biblical times and was introduced to Europe in the mid 1600s.*

PALMS AND BANANAS

Palms, family Palmae, are woody trees or shrubs with fan- or umbrella-like, leafy crowns, whose stems or trunks remain the same in girth for their entire height. Only a few of the nearly 3,000 species occur in temperate regions, the hardiest in general cultivation being *Trachycarpus fortunei*. Its fan-shaped leaves may be 4 feet (1.2m) across and it is native to northern Burma and much of China. Its species name commemorates Robert Fortune (1812-1880). This palm was among the 190 new species he sent back to Europe as living specimens. The seedlings arrived at Kew Gardens near London in 1849, and one of the original plants still grows there.

A similar fan-leaved palm, up to 50 feet (about 15m) tall and often grown as an avenue tree in warm countries, is *Washingtonia filifera*. The genus commemorates George Washington (1732-1799), first President of the U.S., while the species name describes the fibrous apices (tips) of the leaf segments, which give the whole tree a rather wispy appearance. This palm grows wild near springs in the Sonora Desert of southwest North America. The winter resort of Palm Springs, California, is named from the wild specimens that once grew there, and are still numerous, but now deliberately planted.

Bananas (Musaceae) may resemble palm trees, but they are soft herbs. The trunk-like stem is actually composed of overlapping petioles of leaf stalks. True bananas (*Musa* spp.) are grown commercially for their fruit in huge numbers in the tropics, yet rarely as ornamentals, even though their broad, soft green leaves are very decorative.

A hardier species suitable for warmer regions or large glasshouses is the Ethiopian banana, *Ensete ventricosum*. Its single stem dies after it has flowered, a process that takes several years. The plant becomes very large, with a massive bulbous base and leaves up to 20 feet (6m) long, bright green with a red midrib and margin. Some European countries hold competitions to see who can grow the largest specimen. The inflorescence is a pendulous bud of overlapping reddish bracts, which open in sequence to reveal small flowers, whose musky smell and abundant nectar attract bats. The banana-like fruits are filled with large, hard seeds surrounded by small amounts of pulp.

ORIENTAL FAN *The leaves of the Chinese palm* Trachycarpus fortunei *may be 4 feet (over 1m) long.*

A POMPOUS BOTANIST AT OXFORD

UNIVERSITY, ENGLAND, ONCE REMARKED THAT

"PICTURE BOOKS ARE FOR FOOLS, AND ONLY FOOLS USE THEM". HE UNDERESTIMATED THE

ESSENTIAL ROLE OF ILLUSTRATION IN BOTANICAL WORKS AS AN AID TO IDENTIFICATION.

THE FIRST PLANT GUIDE
*A page from Dioscorides'
De Materia Medica.*

THE ART OF ILLUSTRATING PLANTS

Correct identification of plants was of particular importance when the plants were used for their medicinal qualities, as they were until the 1500s. Botanical reference works were dominated until that time by the work of Dioscorides, a Roman army doctor of the first century A.D. Early copies of his *De Materia Medica* such as the *Codex Vindobonensis*, made in A.D.512 for a Byzantine princess, were magnificently and accurately illustrated, but illustration quality deteriorated in later copies.

Stylized illustrations were favored in medieval herbals, until the Italian physician Benedetto Rinio introduced fine naturalistic illustrations by Andrea Amadeo in his *Liber de Simplicibus* (1419). Some of the most sensitive flower drawings of all time were produced by the Renaissance polymath Leonardo da Vinci (1452-1519), while his contemporary Albrecht Dürer (1471-1528) painted a sod of turf (*Das Grosse Rasenstücke*) with exquisite precision in 1503. Over the next 150 years, plants gradually became used and regarded less as medicinal aids, and more for their decorative value, and "florilegia" took the place of the earlier "herbals".

RENEWED EFFORTS
The Renaissance brought a renewed interest in faithful representations of plants such as this drawing of a Mandragora species from Matthioli's Commentarii.

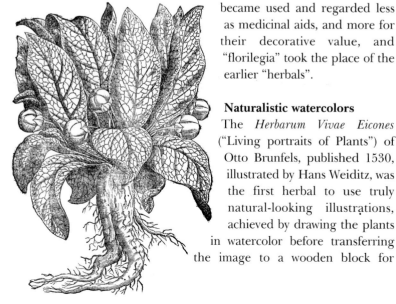

Naturalistic watercolors

The *Herbarum Vivae Eicones* ("Living portraits of Plants") of Otto Brunfels, published 1530, illustrated by Hans Weiditz, was the first herbal to use truly natural-looking illustrations, achieved by drawing the plants in watercolor before transferring the image to a wooden block for printing. Woodcuts of varying quality were used to illustrate works such as John Gerard's *Herball* (1597) and John Parkinson's *Paradisus* (1629), a work in which the transition from medicine to horticulture is almost complete.

By the early 1600s, engraving on copper produced a crisper, more detailed image than the woodcut. Two extraordinary works were produced within a year of each other. The first was Basil Besler's *Hortus Eysttetensis* (1613), a book so immense that it is joked that a wheelbarrow is needed to carry it about. Its engravings record the plants grown in the garden of Konrad, Prince Bishop of Eichstätt, in southern Germany, and furnish (as do other works) a valuable record of the plants then grown and their routes of introduction. The second work is the *Hortus Floridus* of Crispijn van der Pas (1614), a slim volume containing horizontal plates of carefully drawn plants with humorous details such as a mouse nibbling a crocus corm. It was an early coffee-table book, meant to be enjoyed for its pictures and issued in monochrome, with instructions to the reader on how to color the plates.

Scientific accuracy

As botany developed as an independent science, botanical art developed accordingly. The finest botanical artists were trained and supervised by the botanists they served, such as the German Georg Ehret under the patronage of the great Swedish naturalist Carolus Linnaeus.

At the turn of the 18th and 19th centuries, many botanical artists were retained by wealthy patrons to illustrate the plants in their gardens or found on their travels. The French artist Pierre-Joseph Redouté, was employed by

AID TO BOTANY
The roses (above) are by Carolus Clusius from Flanders, an enthusiastic 16th-century botanist. His descriptions of plants and accompanying illustrations proved of enormous help to fellow botanists.

A VASE OF FLOWERS, C. 1688, BY WILLEM VAN AELST.
The fine flower paintings of the Dutch School of the 16th and 17th centuries (right), were beautiful images, but they were not botanically correct.

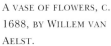

Empress Joséphine Bonaparte to paint her collection of roses and other plants at her favorite garden, La Malmaison (c.1802-1824), near Paris, and a series of illustrations was prepared by Ferdinand Bauer to accompany John Sibthorp's *Flora Graeca* (1806-1840).

Photography has proved no substitute for the detail and sensitivity recorded by a skilled artist, and botanical art continues to thrive as a medium of illustration. Curtis's *Botanical Magazine*, founded in 1787 by William Curtis (1746-1790), provides a continuous link between the botanical artists of today and those of the 1700s, each issue publishing a number of specially commissioned plates of the highest technical accomplishment.

PLANTSMAN'S CHOICE

ZANTEDESCHIA
(ARACEAE)

The Arum family contains some of the oddest plants in cultivation. Their dull-colored inflorescences often give off an appalling smell of carrion—an adaptation to attract pollinating flies—making them an acquired taste for specialist gardeners.

Other, more appealing species, are rainforest epiphytes or climbers. Some of these, such as *Monstera deliciosa*, the "Swiss cheese plant," are favorite house plants, while the waxy

ZANTEDESCHIA
AETHIOPICA
Waxy white spathes bear minute flowers at their center. The plant requires ample moisture to flourish and can be grown at the edge of a pond.

blooms of *Anthurium*, known as "flamingo flowers", are popular for their brilliant shades of pink and red.

The inflorescence of the Arum Lily, *Zantedeschia aethiopica*, is typical of the family. The minute flowers are borne on the yellow spadix, the cylindrical structure found in the center of the broad white spathe. Although it bears the name *aethiopica*, *Zantedeschia* is native only to southern Africa; the anomaly dates to the time when all of Africa was loosely known as Ethiopia. A marsh plant in the wild, *Zantedeschia* favors moist situations in gardens; in many parts of the world it has escaped and become a serious weed in swamps.

As well as the white *Z. aethiopica*, there are several yellow species, some with a dark maroon base to the spathe, and a cultivar of *Z. aethiopica*, with a green (rather than white) spathe, is called 'Green Goddess'. Within the past few years these have been used to breed a new race of miniature hybrids in shades varying from bright yellow to dull red and even violet, which make excellent cut flowers.

BERBERIS
(BERBERIDACEAE)

It is easy to forget that Charles Darwin, the white-bearded evolutionary theorist, was once a young, vigorous field naturalist. Between 1831 and 1836 he took part in a scientific voyage to South America in *H.M.S. Beagle*, under Captain Fitzroy. This gave him ample opportunity to observe and collect specimens of animals, plants, and minerals, as well as gain his first insights into the mechanism of evolution. In 1835, when visiting the Island of Chiloe, off the coast of Chile, Darwin collected a prickly shrub with bright orange flowers, later named *Berberis darwinii* in his honor. One of 450 species of *Berberis* worldwide, *B. darwinii* was introduced to cultivation in Europe by the planthunter William Lobb (1809-63) in 1849. It became a

with aluminum sulfate. Lacecap hydrangeas are derived from the wild species *H. macrophylla* var. *normalis*. All hydrangeas are woodland plants in their natural habitat.

SALVIA
LABIATAE (LAMIACEAE)

Salvia derives its name from the Latin word *salvere* (to heal), because the leaves of the common sage (*Salvia officinalis*) were used medicinally in ancient times. *S. officinalis* is grown for its attractive foliage, as well as for its value as an aromatic kitchen herb. The flowers are a dull purple. Like other culinary herbs in the same family—thyme (*Thymus vulgaris*), rosemary (*Rosmarinus officinalis*), and marjoram (*Origanum spp.*)—sage comes from the fragrant scrubland of the Mediterranean.

Similar habitats in North and Central America also support salvias, which are often shrubs with brightly colored flowers. The crimson *S. microphylla* and the paler *S. greggii* are the parents of a series of shrubs with flowers in shades of yellow, orange, red, pink, and purple. They are superb for gardens in Mediterranean climates, and in California are frequented by hummingbirds. The bedding salvia with spikes of scarlet (now sometimes cream or purple) flowers and bracts is *S. splendens*. It originates from Brazil, where it is a leggy shrub; in cultivation it is usually grown as a compact annual. Flower color in *Salvia* is very diverse: *S. patens*, a tender herbaceous perennial from Mexico has brilliant, electric-blue flowers, while *S. castanea* from the Himalayas and China has dull, red-brown blooms.

parent of *B. x stenophylla*, from which many different clones are cultivated. *Berberis* can be an alternative host for the fungal disease wheat rust, which is an agricultural menace and, as a result, its cultivation is restricted in parts of the U.S.

The evergreen shrub *Mahonia* is closely related to *Berberis*. Also a plant of scrubland and woodland edges, its yellow flowers are usually borne in winter or early spring. *M. Japonica* is a superb, winter-flowering shrub with a fine fragrance, which was introduced from China in 1849 by Robert Fortune.

BERBERIS DARWINII *(above) is a floriferous shrub from South America; its prickly stems make it suitable for hedging.*

HYDRANGEA MACROPHYLLA
(HYDRANGEACEAE)

The German Philipp von Siebold (1791-1866), plant collector and nurseryman, was employed by the Dutch East India Company in Japan as a physician. The influence of this profession gave him privileged access to Japanese gardens, and he built up a remarkable collection of Japanese plants. After being imprisoned for a year for acquiring maps of Japan—normally a capital offence—he was deported in 1830, but was allowed to take his plants with him to Holland. Among them were some of the first hydrangeas to reach Europe, garden varieties of *H. macrophylla* with mopheads of blue or pink bracts. They became very popular as pot plants during the 19th century, but are now usually grown as hardy shrubs.

The precise shade of the bracts is determined by soil acidity and the presence of aluminum in the soil—on chalky soils even the best blue cultivars will turn pink unless treated

HYDRANGEA MACROPHYLLA *cultivars (above) only develop their true colors in acid soil; if lime is present, they become pink.*

SALVIAS *may be small herbaceous plants or robust shrubs with flowers from dull brown, to yellow, bright blue, scarlet, or purple like the* S. guaranitica *(right).*

KNIPHOFIA *(above) often
flowers prolifically after a
bush fire in the wild.*

PASSIFLORA
(PASSIFLORACEAE)

Shortly after the Spanish Conquest of South America, during the 1500s, Jesuit missionaries noticed that a climbing plant bore complex flowers which seemed to symbolize Christ's passion. They called the plant *Passiflora* (passion flower). The outer whorl of ten petals and sepals represented ten of the apostles—Judas Iscariot and Doubting Thomas were excluded—while a whorl of narrow segments represented the multitude of Christians worldwide. Its central column bears both anthers and styles—the anthers represented the five wounds Christ suffered, and the stigmata the three nails of the crucifixion, while the globular ovary suggested the hammer used to bang in the nails. The Jesuits used the flower to explain their religious teachings to the Amazonian Indians.

There are many species of *Passiflora*, all indigenous to the New World and all bearing similarly complex flowers. *P. incarnata*, from the southeastern U.S., reached Spain some time before 1525. *P. caerulea*, from Brazil, has creamy-white petals and a crown of dark blue segments which are dull red at the base. It is one of the hardiest species and can be grown outdoors in milder places in temperate regions. Most other species need tropical conditions, but their magnificent blue, purple, or scarlet flowers, often followed by attractive fruit, make greenhouse cultivation worthwhile. The commercial passion fruit, or grenadilla, comes from *P. edulis*.

KNIPHOFIA
(LILIACEAE)

The bright red spikes of many species of *Kniphofia* have given the plants the common name of red hot poker, while their generic name commemorates Johannes Hieronymus Kniphof (1704-1763), a German professor of medicine and an amateur botanist. He invented a method of plant illustration which involved recording the outlines by pressing dried specimens covered with ink onto a sheet of paper, and in 1762 produced a large book of illustrations using this technique. It included an impression of a plant he called *Aloe uvaria*, which was soon renamed *Kniphofia uvaria* in his honor. *K. uvaria* is common in South Africa, and was the first red hot poker to reach Europe, where it became a parent of many hybrids. *Kniphofias* are found on the mountains of Africa from the Cape of Good Hope to Ethiopia, with one species each in Madagascar and southern Arabia. They relish damp ground, and mountain swamps can be covered in thousands of them. Flowering is especially prolific after a fire, which burns off dead vegetation and releases nutrients into the soil.

The robust *K. caulescens* grows from a short stout trunk. A clone brought from South Africa by Ken Burras, former superintendent of the Oxford Botanic Garden, has almost blue foliage and produces its flowers in late fall—the last large plant to flower before winter sets in. Other species, such as *K. triangularis*, are much more slender and elegant; breeders have used them to create miniature red hot pokers that are better suited to small gardens.

PASSIFLORA
*The complex structure of
the passion flower (right)
suggested the instruments
of Christ's Passion.*

PHLOX
(POLEMONIACEAE)

The woods, meadows, and mountains of North America are the native habitat of the 67 species of *Phlox*. Their five-lobed flowers caught the eye of John Tradescant the Younger on his journeys to America in the 1630s, but none were successfully introduced to Europe until the early 18th century, when the American botanist John Bartram sent plants to his correspondent Peter Collinson in London. Among the collection was a clump of the rock-garden plant *P. subulata*, now available in many shades of white, pink, and purple which produce a dazzling display in early summer.

Phlox paniculata, a robust plant from the woodland edges of eastern North America, was another of the species sent by Bartram to Collinson. Pink or purple in the wild, breeders have selected cultivars with flowers in shades of white to bright red, sometimes even bicolored. Their stiff stems, rounded inflorescences, and appealing scent are a feature of late summer herbaceous borders. They are extremely hardy and thus particularly popular in Siberia.

Phlox drummondii from Texas is an annual species often grown as a bedding plant for its heads of multi-colored flowers. Originally reddish purple, these are now available in all shades of pink, red, and purple, as well as pastel pinks, oranges, and even yellow. The Mexican *P. nana* has even brighter flowers in brilliant gold and flaming tangerine, but it is not easy to cultivate successfully in colder climates.

HELLEBORUS
(RANUNCULACEAE)

The flowers of hellebores range from green to almost black. Natives of Europe from as far east as the Caucasus, with a solitary species in China (which bear large blooms in midwinter), they are robust hardy plants, found in subalpine meadows and open woodlands. *Helleborus argutifolius* (syn. *H. corsicus*) is found only on the Balearic islands in the western Mediterranean; it has long stems crowned by leathery leaves below soft green flowers; an architectural plant that starts to flower in midwinter in northern gardens. The white flowers of the "Christmas rose," *H. niger*, belie its specific name which means black. This is derived from the color of the roots which were used in medieval medicine. It comes from the Alps, and although an early-flowerer, is seldom open until early spring in temperate regions.

The many color forms of *H. x hybridus* are derived from the Caucasian *H. orientalis*, which has white or pink flowers. *H. x hybridus* flowers in early spring before its thick palmate leaves unfurl; it is sometimes known as the "Lenten rose." Although typically white, pink, or dull red, sometimes with deeper spots inside, breeders have created plants with butter-yellow or slaty-blue flowers. Others have a picotee edge, and doubles have been produced recently. The hellebores of premier breeder, Helen Ballard (d. 1996) of Worcestershire, England, are regarded as the finest ever raised and sell for high prices. Her work is being continued and extended by many other breeders in Europe and North America.

HELLEBORUS X HYBRIDUS
The pendulous flowers appear in late winter and may be exquisitely marked inside.

PHLOX PANICULATA
The wild form has smaller flowers and a more graceful appearance than the larger garden derivatives (left).

CEONOTHUS
The blue or mauve flowers have inspired the name "California lilac." Most species do best in frost-free climates.

CEANOTHUS
(RHAMNACEAE)

Ceonothus americanus was discovered in the early 1700s by the naturalist Mark Catesby (1669-1749), whose *Natural History of Carolina* was to become an American classic. Its leaves were widely used for tea during the American Civil War, and an extract from the bark helped to staunch wounds. The plant comes from the genus *Ceanothus*, which includes about 60 species found in North America and Mexico. They are most abundant in California, where there are 45 species, giving them the common name of "California lilac". Most of these are important constituents of the chaparral scrubland. *C. americanus* is a deciduous species with dull whitish flowers; it is superseded in horticultural value by many of the evergreen Californian species and their hybrids, which produce bright blue flowers in late spring. Unfortunately, many of these are very frost-sensitive and need a warm place to survive. A few are hardy to 50°F (-15°C), including the northern Californian *C. thyrsiflorus* var. *repens*, which makes a sprawling mound covered in dark blue flowers.

HEBE
(SCROPHULARIACEAE)

Allan Cunningham (1791-1839) was one of the first collectors from Kew to explore the horticultural potential of Australia and New Zealand, where he spent the years 1814-31. During this period an enormous number of seeds and bulbs were sent to Kew, including those of the first species of *Hebe* to

NICOTIANA X
SANDERAE *(above right) is the annual "tobacco plant" used for bedding display in summer.*

HEBE SPECIOSA *was one of the first species of this New Zealand genus to reach Europe; it grows well in maritime areas.*

reach Europe. Initially, they were included in the related genus *Veronica* on account of their similar flowers. *Hebe*—whose name means "youth" in Greek—is a genus of small shrubs with evergreen leaves. It is chiefly found in New Zealand, but a few species also occur in Australia and South America. The species found by Cunningham were *H. speciosa* and *H. salicifolia*, both leathery-leaved shrubs with attractive flowers, deep purple and white or mauve respectively. *H. speciosa* often has purple-tinged leaves, which makes it a very handsome plant. Although not frost-hardy, it is extremely tolerant of wind and salt spray, which makes it (and many other hebes) a very useful plant for mild coastal areas. Many hybrids have been raised using these two parents (as well as others); some are hardy to 50°F (-15°C), but most are best in milder climates.

NICOTIANA
(SOLANACEAE)

Native Americans had been inhaling smoke from the smoldering leaves of a tall white-flowered annual plant for centuries by the time Columbus landed in Cuba in 1492. His crew was astonished to see people apparently breathing smoke. The local name for the plant was *tabaco*. *Nicotiana*, the scientific name, commemorates the Frenchman Jean Nicot de Villemain, who introduced the plant to France in 1560. At first, *Nicotiana tabacum* was grown in Europe as an ornamental for medicinal purposes (the nicotine it contains is a useful insecticide), but by the early 1600s people were beginning to smoke it as a relaxant.

The English courtier Sir Walter Raleigh was one of the first to take up smoking seriously. It is said that a servant, unaccustomed to his ways and believing him to be on fire, once doused him with a bucket of water. King James I was an enemy of both Raleigh and tobacco, writing the fiercely worded *A Counterblast to Tobacco* in which he recognized the dangers it poses to health. He had Raleigh beheaded, but was less successful in preventing smoking.

N. tabacum is seldom grown as an ornamental, but *N. sylvestris*, which reaches 6 feet (1.8m) high and bears many long white flowers, is very similar. Its flowers open in the evening, attracting moths by their scent, as do those of the smaller *N. x sanderae*, the common "tobacco plant" used for summer bedding. Its flowers may be

purple, red, white, or even lime green. The elegant *N. langsdorfii* has pendulous, small, green flowers with anthers that burst open to reveal bright blue pollen, while *N. glauca* is a small tree with glaucous leaves and sprays of lime green flowers.

TROPAEOLUM
(TROPAEOLACEAE)

The genus *Tropaeolum* derives its name from the stand set up by the Roman army after a battle on which were displayed the shields and helmets of the vanquished. Seeing the rounded leaves and helmet-shaped, orange flowers of *T. majus*, the Swedish botanist Carolus Linnaeus adapted the name of this ancient symbol for the genus of climbing and sprawling plants from South America. The English name "nasturtium" is also of Classical derivation, being the old name for watercress (*Rorippa nasturtium-aquaticum*), whose tingling flavor is shared by *Tropaeolum*. The leaves and flowers are said to be extremely rich

in vitamin C, but can be an acquired taste. The small-flowered *T. minus* was introduced to Europe in the 1500s, but died out. It was later superseded by *T. majus*, the common annual nasturtium grown today. This species originates from Peru and is typically a climbing plant, although non-climbing cultivars have developed in cultivation and are known as Tom Thumb nasturtiums. A variegated cultivar, 'Alaska', which has leaves that are heavily splashed with creamy white, is interesting as one of the few variegated plants to produce variegated offspring from seed. More graceful is *T. peregrinum*, the "canary creeper", whose small yellow flowers have a decided resemblance to a flying bird. It is also an annual, but others are perennials, including *T. speciosum*, the "flame flower". This beautiful plant has bright red flowers produced from thin stems which scramble up through shrubs; it does extremely well in cool moist conditions and often covers large lengths of hedge in brilliant color.

TROPAEOLUM MAJUS, *an easily grown annual (below), flowers more prolifically and produces fewer leaves when grown in poor conditions.*

GLOSSARY

alpine: Low-growing plant from mountains above tree-line, or tundra.

anther: Flower's male reproductive organ, producing and releasing pollen.

annual: plant that grows, flowers, and dies within one year.

axil: the angle between leaf and stem.

biennial: a plant that flowers in the second year of growth, and then dies.

bract: modified, leaf-like structure beneath a flower, usually smaller than a true leaf.

bulb: subterranean storage organ consisting of a very compact stem surrounded by swollen leaf bases in which food reserves are stored.

calyx: outer whorl of floral parts, usually green and not very showy.

carpel: entire female reproductive organ of a flower, composed of stigma, style, ovary, and ovules, usually situated at the center of the flower.

chromosome: a body found in the cell's nucleus that contains the genetic material DNA. Cells possess a definite number of chromosomes, which varies according to the species. In the normal state of an organism, chromosomes are present as two sets in every cell. This is known as the diploid number. Haploid cells with only one set of chromosomes occur during reproduction in pollen grains and ovules; fertilization restores the diploid state. If the chromosome number is doubled further, to four sets, the plant is known as tetraploid, and will usually show increased vigor and size, as it will if triploid (3 sets of chromosomes). Triploid plants are usually sterile.

clone: genetically identical group of plants derived from vegetative propagation from a single parent, using techniques such as division, cuttings, or grafting.

corm: subterranean storage organ consisting of a solid stem. It is replaced annually from the base of the current season's growth; the old corm dies, as in Crocus, Gladiolus.

corolla: an inner whorl of attractive floral parts—the petals—but especially used when these are united into a single, tubular or bell-shaped structure, when the free tips are known as corolla-lobes.

cultivar: distinct plant variety that has arisen in cultivation by hybridization or selection. Usually a vegetatively propagated clone, but may be raised from seed if the progeny are uniform within narrow limits of variation. Cultivars are denoted by single quotation marks, as in Rosa 'Peace'.

diploid: *see chromosomes.*

DNA: Deoxyribonucleic acid, the genetic material of organisms, contained as tight spiral molecules within the chromosomes; the sequence of paired bases along its length constitute the genetic code.

epiphyte: a plant that uses another plant for support, but without parasitizing it (although its presence may eventually damage the support).

F1: first generation progeny from a cross of two distinct parents. Usually applied to progeny of crosses between pure-bred cultivars, whose appearance and performance will be uniform. Seedlings (the F2 generation) from seed saved from F1 plants will show great variability and desirable characteristics will probably be lost.

family: a taxonomic unit containing related genera. A family name conventionally ends in -aceae, but a few important families end in -ae, as in Compositae (also known as Asteraceae).

form: (strictly forma) a very minor variant of a wild plant of little botanical significance, but sometimes horticulturally desirable, as in *Cyclamen hederifolium forma albiflorum*, the white-flowered form of the usually pink-flowered species. Abbreviated to f.

fruit: any seed-bearing organ, including edible fleshy fruit, dry pods and capsules.

fynbos: the species-rich, shrub-dominated vegetation from the Cape region of South Africa.

garigue: scrub vegetation of low, scattered, often spiny and heath-like shrubs with bare ground between, typical of mediterranean regions.

genus: taxonomic unit containing one or more species of closely related organisms; related genera (plural) make up a Family. A generic name is always written in italics with an initial capital e.g. *Cyclamen*.

glaucous: colored bluish-gray or whitish, due to a waxy bloom that covers the plant's surface (especially the leaves).

haploid: see chromosomes.

herbaceous: in gardening as opposed to botanical terms, describes plants which die down every year but regrow the next; hence herbaceous border for a planting scheme made up of such plants. In botany refers to plants which have soft, leafy rather than permanent woody, stems such as petunias.

Ray floret

Disc floret

Bract for single floret

Composite head

Bracts of involucre

Capitulum in section

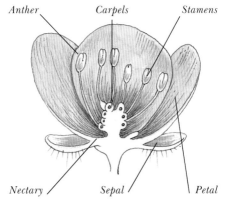

Anther　　　*Carpels*　　　*Stamens*

Nectary　　　*Sepal*　　　*Petal*

Ranunculaceous flower

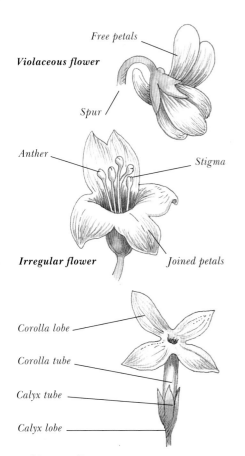

Violaceous flower

Free petals

Spur

Anther

Stigma

Irregular flower

Joined petals

Corolla lobe

Corolla tube

Calyx tube

Calyx lobe

Rubiaceous flower

hybrid: the progeny of two different parents, usually two species, but also between distinct clones. The process of hybridization is also termed "crossing." Hybrids occur infrequently in wild populations, but are common in gardens where artificial pollination is easily carried out.

inflorescence: structure on which the flowers are borne, usually distinct from the vegetative parts of the plant.

native: a plant which occurs naturally in an area and has not been introduced from anywhere else.

naturalized: plants that have become absolutely established in an area away from their original region.

nectary: nectar-producing organ.

ovary: the lower part of the female organ of a flower, containing the ovules (embryonic seeds) in which they develop into seeds. If located above the calyx the ovary is termed superior (e.g. Liliaceae), if below the calyx, inferior (e.g. Amaryllidaceae).

ovule: female embryo; when fertilized by pollen it develops into a seed.

parasite: plant which grows on or within another drawing nourishment from it.

petal: a floral organ, placed in a whorl between calyx and stamens.

perennial: a plant that survives and flowers for several years in succession. The stems of a herbaceous perennial are annual, dying after flowering (e.g. *Aster novae-angliae*), but are replaced from the rootstock each year.

pollen: the male reproductive cells, produced from the anther.

rhizome: thickened, subterranean stem, usually growing horizontally, as in *Iris germanica*.

rootstock: the point from which roots and shoots arise, especially in a herbaceous perennial.

rosette: cluster of leaves apparently radiating from the same point.

sepal: individual division of the outer whorl of floral parts (calyx).

seed: mature dispersal unit of flowering plants, derived from the fertilization of an ovule by a pollen grain.

selection: process whereby favorable genetic variation is recognized and perpetuated by the more successful reproduction of organisms that show it. Natural selection is the force by which evolution proceeds; artificial selection is the more rapid process undertaken by plant- and animal-breeders in the human environment.

shrub: woody plant branching from the base, usually less than 15 feet (4.5m) in height.

spathe: large bract enclosing a flowerhead, as in *Arum*.

species: the basic unit of biological classification; a group of individuals with shared characteristics different from those of other similar groups, that can interbreed freely within that group, but not (or less freely) with a different group. Specific names are written in italics with a lower-case first letter, and

as in these examples of specific names from the genus *Cyclamen*, are usually descriptive (e.g. *hederifolium*, ivy-leaved), geographical (*graecum*, Greek) or commemorative (*rohlfsianum*, of Gerhard Rohlfs, the botanist who discovered this species). Abbreviated to sp. with spp. as the plural form.

stamen: entire male organ of a flower, composed of the anther and its supporting filament, are usually found in a whorl above the petals.

stigma: the pollen-receptive apex of the style.

style: the usually elongated part of the carpel that bears the stigma.

subspecies: a division of a species based on minor morphological distinctions, often coupled with differences in its geographical range or ecological requirements. Usually abbreviated to subsp. or ssp.

succulent: a fleshy plant, storing water in its leaves or stems, usually from arid regions.

triploid: see chromosomes.

tree: a woody plant with (usually) a single stem that branches at some distance above the base, reaching 15 feet (4.5 m) or more in height

tuber: a perennial subterranean storage organ consisting of swollen stems e.g. *Cyclamen*. The term is often also used for tuberous roots—storage organs derived from a plant's roots, as in Dahlia.

variety: (strictly, *varietas*) a lower subdivision of a species, less strongly distinct than a subspecies. Correctly used for populations of wild variants, as in the tetraploid *Cyclamen hederifolium* var. *confusum*; cultivar is an equivalent for plants that arose in cultivation and which are usually more uniform than those in a wild population. Abbreviated to var.

Ovary superior *Ovary superior* *Ovary inferior*

INDEX

CREDITS

Quarto would like to acknowledge and thank the following for providing pictures reproduced on the following pages in this book

AGK London: 60 ñ61, 175 br; British Library, London: 8l; Neil Davies: 47 tl, 47 c; Garden Picture Library:8 r, 9 l, 18, 18b,19 c, 19b, 20 br, 21 23b, 24-25, 26 tl, 26 ñ 27, 27 br, 28, 39 tol, 39 tr, 39 bl, 41 bl, 42, 43, 44, 45t, 45b, 47t, 49t, 50, 51, 74 tl, 74 ñ 75,

75b,76bl, 76 cr, 77bl, 77r, 79, 80, 83t, 114 t & b, 115, 116, 117, 118, 120 t & b, 121, 132 t & bl, 136 r, 141, 152t & br, 153 tl,153tr, 153 br, 155, 156, 157, 158-9, 160 bl & br, 161 bl, 162, 163, 164 tl & bl, 164-5, 168, 170 ñ 171, 172 t, 173t,, c & b, 174, 176 t & b, 177, 178 tl, 178 tr, 178 tl, tr, cl, & bl; 179 l, 180, 183,184 b, 184 tl, 185 tr, 185 br, 186 b, 188, 194, 196, 200 t & b, 202 t & b, 203, 206, 210 t & B, 210 ñ211; (Garden Picture Library photographers are: Philippe Bonduel, Chris

Burrows, Rex Butcher, Brian Carter, Bob Challinor, Densey Clyne, Dennis Davis, Ron Evans, John Fairweather, John Ferro Sims, Vaughan Fleming, Nigel Francis, Jon Glover, Sunniva Harte, Neil Holmes, Jacqui Hurst, Michele Lamontagne, Zara McCalmont, Mayer/Le Scanff, John Miller, Martine Mouchy, Jerry Pavia, Peter Phipp, Morley Read, Howard Rice, Kate Zari Roberts, David Russell, J S Sira, Brigitte Thomas, Michel Viard, Juliette Wade,

Micky White, Steve Wooster). J. Grimshaw: 10, 11, 47 tl, 47 c, 167, 182, 90 b, 90 ñ 91, 192b, 193 tl, bl & r, 215 t; Clive Nichols Garden Pictures: 214 t Jane Nichols; Jerry Pavia: 148-9, 172 bl, 172 br, 179 r, 181, 213x, 214 b, 215 b, 216b, 217 t; Harry Smith Collection: 18 t, 19t, 20 t, 20, 23 t, 38 t, 39 br, 40-41, 46 t, c & b, 49b, 60 tl, 76 t, 83, 132 c, 132 br, 133 t & r, 134, 135, 136l, 138, 139, 152 bl, 154, 156 tl, 159, 160 t & c,184 t, 186 tl, 197, 198, 200c, 201t, 204, 212, 213 t, 213 r, 216 t.

The picture on the following pages are the copyright of Quarto: 29, 78-9, 81, 161 tl & cr, 166l & r, 178 br, 189, 192t, 195 r, 201 bl, 215, 217b

Key
t = top
b = bottom
c = centre
l = left
r = right